One-Sample z Statistic (Sec. 4.B)

$$z = \frac{\bar{x} - \mu_0}{s/\sqrt{n}}$$

(4.1)

One-Sample t Statistic (Sec. 4.B)

$$t = \frac{\bar{x} - \mu_0}{s/\sqrt{n}}$$

(4.2)

$$df = n - 1$$

Two-Sample z Statistic (Sec. 4.C)

$$z = \frac{\bar{x}_A - \bar{x}_B}{\sqrt{\dfrac{s_A^2}{n_A} + \dfrac{s_B^2}{n_B}}}$$

(4.3)

Two-Sample t Statistic (Sec. 4.C)

$$t = \frac{\bar{x}_A - \bar{x}_B}{\sqrt{\dfrac{\Sigma(x_A - \bar{x}_A)^2 + \Sigma(x_B - \bar{x}_B)^2}{n_A + n_B - 2}\left(\dfrac{1}{n_A} + \dfrac{1}{n_B}\right)}}$$

(4.4)

$$df = n_A + n_B - 2$$

Sum of Squared Deviations (Sec. 4.C)

$$\Sigma(x_A - \bar{x}_A)^2 = (n_A - 1)s_A^2$$

(4.5)

Paired-Sample t Statistic (Sec. 4.D)

$$t = \frac{\bar{d}\sqrt{n}}{s_d}$$

(4.6)

$$df = n - 1$$

One-Sample z Statistic for Proportion (Sec. 4.E)

$$z = \frac{\hat{p} - p_0}{\sqrt{\dfrac{p_0(1 - p_0)}{n}}}$$

(4.7)

Two-Sample z Statistic for Proportions (Sec. 4.E)

$$z = \frac{\hat{p}_A - \hat{p}_B}{\sqrt{\hat{p}(1 - \hat{p})\left(\dfrac{1}{n_A} + \dfrac{1}{n_B}\right)}}$$

(4.8)

$$\hat{p} = \frac{n_A\hat{p}_A + n_B\hat{p}_B}{n_A + n_B}$$

(continued on back cover)

ESSENTIALS OF
STATISTICS

STEPHEN A. BOOK *California State College, Dominguez Hills*

ESSENTIALS OF STATISTICS

McGRAW-HILL BOOK COMPANY

*New York St. Louis San Francisco Auckland Bogotá Düsseldorf
Johannesburg London Madrid Mexico Montreal New Delhi
Panama Paris São Paulo Singapore Sydney Tokyo Toronto*

This book was set in Helvetica by Black Dot, Inc.
The editors were A. Anthony Arthur, Janice Lynn Rogers,
and James. W. Bradley;
the designer was Joan E. O'Connor;
the production supervisor was Leroy A. Young.
R. R. Donnelly & Sons Company was printer and binder.

ESSENTIALS OF
STATISTICS

1234567890DODO783210987

Library of Congress Cataloging in Publication Data

Book, Stephen A
   Essentials of statistics.

   Abridgement of the author's Statistics.
   Includes index.
   1. Statistics. I. Book, Stephen A. Statistics.
II. Title.
HA29.B7332   519.5   77-6787
ISBN 0-07-006464-4

To the memory of my uncles

CONTENTS

PREFACE

By presenting applied examples and exercises to introduce the most basic statistical methods, this one-quarter or one-semester text illustrates the role of statistics in the solution of applied problems. All statistical methods are fully explained and illustrated in terms of situations drawn from economics and the managerial sciences, psychology, sociology, political science, and the other social sciences, and the life, medical, and agricultural sciences. Through the examples solved in the text, discussed by the instructor in the classroom, and worked by the student outside of class, the student will develop an understanding of the common uses of statistical reasoning in answering quantitative questions in all the above fields of knowledge.

In order to understand the applied examples and exercises, it is not necessary for the student to begin the course with a formal background in mathematics. While some experience in algebra would allow the student to progress faster, the text begins at a very low level, and new mathematical techniques are introduced very slowly as needed. If the instructor provides a few minutes of appropriate (depending on the educational backgrounds of the students) remarks on topics in arithmetic and algebra from time to time, the student should be able to build up the skills necessary to handle each succeeding section.

The heart of the elementary statistics course consists of the first five chapters. In these chapters, the student will learn graphical and numerical ways of describing a set of data, the binomial and normal distributions, confidence intervals, fundamentals of hypothesis testing, and linear regression and correlation. Depending on the interests of the instructor and the students, there will often also be time to discuss one or more of the special topics covered in later chapters: probability, the chi-square test of independence, analysis of variance, and nonparametric statistics.

This text has benefited greatly from the comments and suggestions of Professors James M. Edmondson of Santa Barbara City College, Walter H. McCurdy of Bradley University, Linda Mercier of the State University of New York Agricultural and Technical College at Delhi, Charles O. Minnich of Mt. San Antonio College, John S. Mowbray of Shippensburg State College, Frederic C. Parker of Jefferson Community College, and Douglas A. Zahn of Florida State University. To all these professors, I wish to express my appreciation and gratitude for their detailed reading of the manuscript and their valuable suggestions for improvement.

I would also like to thank Mr. Thomas A. Cochran of the Mathematics Department of Carson High School, Carson, California, for his assistance in proofreading the original manuscript and in working out the solutions to all the exercises.

I would like to thank the editors, Tony Arthur, Alice Macnow, and Jim Bradley of McGraw-Hill, for their expert guidance during the course of this project.

I am grateful to the Hafner Press and to the *Biometrika* Trustees for their permission to reproduce certain statistical tables in the Appendix.

Finally, I will close by acknowledging the valuable assistance of my wife, who was a history major in college, in reading the manuscript from a nonmathematician's point of view.

Stephen A. Book

ESSENTIALS OF
STATISTICS

ORGANIZING STATISTICAL DATA

Statistics, an applied branch of mathematics, has as its goal the organization, analysis, and explanation of facts and figures arising out of the study of the social, behavioral, managerial, and natural sciences. Using statistics, we seek to develop general rules of behavior of biological, economic, and psychological phenomena, for example, for the dual purposes of understanding these phenomena as they are and of controlling and improving them for society's benefit. Statistics originated almost a century ago in attempts to understand the factors influencing industrial and agricultural production, in order to better control these factors and so to increase production at reasonable cost. Today the aim of statistics is no different, although it has expanded its vista to include additional aspects of the various sciences.

A statistical problem begins with a question of applied interest under consideration in one of the various sciences mentioned above. Mathematical techniques of statistical analysis enter the discussion when some data bearing on the problem are accumulated. (The proper methods of accumulating data are studied in a relatively nonmathematical branch of statistics called "statistical

TABLE 1.1
Demand for electricity over 100 days
(Millions of kilowatthours)

25	32	28	26	30	21	36	27	32	32
33	23	31	29	33	31	29	33	28	31
27	29	29	33	27	28	32	23	30	25
29	32	21	24	30	31	28	32	31	34
31	26	30	30	34	26	26	29	32	22
24	31	28	32	28	37	30	35	27	33
28	30	33	30	31	30	32	28	31	28
29	26	32	24	29	32	30	22	34	26
32	27	27	34	29	31	33	39	30	33
30	28	31	30	37	31	25	32	33	28

sampling.") After the data are collected, the next step is often the organization of the data in an understandable and easily communicable format. Sometimes the data can be organized into a pictorial or graphical structure so that it can be understood by persons not very familiar by training or experience with numerical descriptions. Visual methods, when available, are excellent vehicles for communicating statistical information quickly and clearly to those not trained in numerical thinking.

To persons accustomed to mathematical thinking,[1] however, numerical descriptions are infinitely more useful. As we will discover in future chapters of this text, numerical descriptions communicate a wealth of detailed and precise information that belies the relative simplicity of the methods involved. In this chapter, we introduce some of the important techniques of organizing a set of data so as to more easily reveal its important patterns and characteristics and convey the information it contains.

SECTION 1.A
FREQUENCY DISTRIBUTIONS AND THEIR GRAPHS

An electric utility must anticipate the demand for electricity in the region it serves in order to be sure of being able to meet normal demands. To get some idea of the demand for electricity, a major electric company recorded the demands upon it each day for the past 100 days. Table 1.1 contains the data obtained by the electric company: the demand for electric power, in millions of kilowatthours, over the past 100 days.

As can immediately be seen from Table 1.1, not much can be specified about the demand for electricity from a glance at the data. The problem is that we have an unorganized collection of numbers, seemingly listed in random fashion and conveying no clear information. No trends indicating possible de-

---

[1]The objective of this book is to make you one of them!

TABLE 1.2
Frequency distribution of demand for electricity

Interval, millions of kilowatthours	Number of Days Having Demand in the Interval	
20–22	||||	4
23–25	ЖЖ |||	8
26–28	ЖЖ ЖЖ ЖЖ ЖЖ |||	23
29–31	ЖЖ ЖЖ ЖЖ ЖЖ ЖЖ ЖЖ ||||	34
32–34	ЖЖ ЖЖ ЖЖ ЖЖ ЖЖ |	26
35–37	||||	4
38–40	|	1
		Sum = 100

mand levels are discernible. In order to make use of the mass of available data, we must organize it into an understandable form.

We first divide the range of data into a reasonable number of intervals and count the number of data points falling into each interval. This procedure gives us what is called a "frequency distribution" of the data and is illustrated in Table 1.2. Because the minimum daily demand for electricity is at the level of 21 million kilowatthours over the 100-day period, while the maximum daily demand is at 39 million kilowatthours, a reasonable set of intervals to use would be 20 to 22, 23 to 25, 26 to 28, 29 to 31, 32 to 34, 35 to 37, and 38 to 40. In choosing a set of intervals, care must be taken in regard to the following points:

1. Every data point must fall into *exactly one* of the intervals, no more and no less.
2. The number of intervals must not be too large, for then no substantial improvement in understanding the trend of the data would be made.
3. The number of intervals must not be too small, for then variations among the data points would be obscured.
4. It is not necessary that all intervals have the same length, but extreme fluctuations in sizes of consecutive intervals should be avoided.

Having agreed upon the intervals, the next step is to tally the number of data points falling in each interval. Basically, this is all there is to the task of constructing a frequency distribution. The result appears in Table 1.2. As a check on the correctness of our counting, we can sum the number of days in the far right-hand column. If our counting was correct, we should get a total of 100 days, the number of days involved in the electric company's data.

Some useful information is apparent from a glance at the frequency distribution of Table 1.2, whereas almost nothing could be learned from a glance at the original data of Table 1.1. For example, two things we can learn from the frequency distribution are the following:

TABLE 1.3
Frequency distribution of demand for electricity, modified for construction of histogram

Original Interval	Modified Interval Boundaries	Number of Days
20–22	19.5–22.5	4
23–25	22.5–25.5	8
26–28	25.5–28.5	23
29–31	28.5–31.5	34
32–34	31.5–34.5	26
35–37	34.5–37.5	4
38–40	37.5–40.5	1

1. Approximately one-third of the time (34 days out of the 100) the demand is about 30 million kilowatthours.
2. The demand exceeds 34 million kilowatthours only about 5% of the time (the 4 days corresponding to the interval 35 to 37, and the 1 day corresponding to the interval 38 to 40).

Since these two facts contain *specific* information, they are much more useful than the mere collection of numbers in Table 1.1.

There are two common vehicles for pictorially communicating the information in a frequency distribution. The first of these methods is the "histogram" (or bar graph), and the second is the "frequency polygon" (or line graph). While the two types of graphs are closely related, there are important differences in the details of constructing them.

THE HISTOGRAM

To construct a histogram, we have to modify the frequency distribution in such a way that there are no gaps between intervals. For the data in Table 1.2, we close the gap between the two intervals 20 to 22 and 23 to 25 by setting the point of division at 22.5, halfway between 22 (the end of the first interval) and 23 (the beginning of the second). To keep the lengths of all the intervals the same, we consider that the first interval starts at 19.5 and ends at 22.5, while the second starts at 22.5 and ends at 25.5. The modified frequency distribution appears in Table 1.3. Using Table 1.3, we construct the histogram of Fig. 1.1 by drawing bars one interval wide to a height corresponding to the number of days in that interval. The first bar, for example, is 4 units high.

THE FREQUENCY POLYGON

To construct a frequency polygon, we have to modify the original frequency distribution (Table 1.2) in a different manner. We have to find the midpoint of

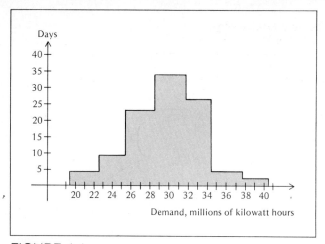

FIGURE 1.1
Histogram of demand for electricity.

each of the intervals, for we now consider each interval to be represented by its midpoint. From Table 1.2, we determine the midpoints of the intervals preceding 20 to 22 and following 38 to 40, as if those intervals really existed in the table. The frequency distribution modified in preparation for the frequency polygon appears in Table 1.4. The frequency polygon appears in Fig. 1.2. The frequency polygon is constructed by connecting dots[1] which represent the number of days that each demand level is attained, considering an interval of demand levels to be represented by its midpoint.

You should observe that the histogram and the frequency polygon both communicate the same information, namely, the facts contained in the frequency distribution of Table 1.2. They both illustrate that demand for electricity

[1]As in the child's game follow-the-dots.

TABLE 1.4
Frequency distribution of demand for electricity, modified for construction of frequency polygon

Interval	Midpoint	Number of Days
(17–19)	18	0
20–22	21	4
23–25	24	8
26–28	27	23
29–31	30	34
32–34	33	26
35–37	36	4
38–40	39	1
(41–43)	42	0

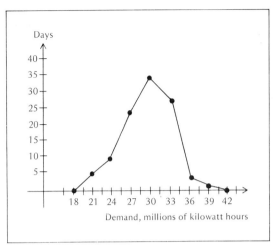

FIGURE 1.2
Frequency polygon of demand for electricity.

~~is rarely below 24 million kilowatthours, very often is between 26 and 34 million kilowatthours, and falls off sharply above 35 million kilowatthours.~~ (These facts were not at all evident from the original data in Table 1.1.)

EXERCISES 1.A

1   A geographer studying West Africa would like to know the size distribution of stones in the delta region where the Niger River runs into the Gulf of Guinea. He collects 50 stones and measures the diameter of each, and so obtains the following measurements, in centimeters:

4	5	9	13	8	19	9	8
17	3	6	7	10	14	20	1
6	11	2	7	12	17	2	18
3	7	14	1	8	3	12	5
6	16	8	11	4	7	6	5
4	6	10	7	12	7	6	15
5	9						

(a)   Construct a frequency distribution of the diameters based on the intervals 1 to 4, 5 to 8, 9 to 12, 13 to 16, and 17 to 20.
(b)   Draw a histogram which illustrates the frequency distribution of part a.
(c)   Draw a frequency polygon which illustrates the frequency distribution of part a.

2   A popular index of common stocks traded on the New York Stock Exchange shows that current values, in dollars, of the 500 stocks listed have the following frequency distribution:

Dollar Value	No. of Stocks Having Value in Interval
0.00– 9.99	51
10.00–19.99	98
20.00–29.99	101
30.00–39.99	151
40.00–49.99	48
50.00–59.99	25
60.00–69.99	13
70.00–79.99	8
80.00–89.99	4
90.00–99.99	1

Construct a frequency polygon based on the frequency distribution of stock prices.

3 The 50 employees of a small factory have yearly earnings distributed as follows:

Earnings Level, dollars	No. of Employees at That Level
0.00– 2999.99	3
3000.00– 5999.99	5
6000.00– 8999.99	18
9000.00–11,999.99	15
12,000.00–14,999.99	6
15,000.00–17,999.99	0
18,000.00–20,999.99	3

Draw a frequency polygon which illustrates the data.

4 One hundred housing units in a major U.S. city are selected at random from the official property tax list. Each is rated according to quantity and quality of plumbing, electrical, and other mechanical facilities. Each unit is assigned a score from 0 to 100, and the resulting ratings appear in the following frequency distribution:

Score	No. of Housing Units
0.0– 19.5	30
19.5– 39.5	15
39.5– 59.5	5
59.5– 79.5	10
79.5–100.0	40

(a) Draw a histogram which illustrates the data.
(b) Construct a frequency polygon based on the data.

5 A psychologist conducted a learning experiment aimed at finding out how quickly monkeys were able to learn that a red light means "stop," while a green one means "go." Each time the monkey made the correct decision, he

or she was rewarded with a banana, and each time the monkey made a wrong response, he or she was punished with an electric shock. One hundred monkeys participated in the experiment, and the following data lists the number of trials necessary for each to learn the lesson:

1	6	17	11	4	18	24	34	2	26
4	2	7	3	12	5	10	1	15	3
16	5	3	8	9	13	1	20	2	14
6	2	8	4	9	5	14	1	2	10
1	7	7	6	5	10	5	9	15	3
6	25	3	8	8	1	6	13	19	10
20	2	7	12	4	9	2	7	9	18
1	6	11	3	19	5	8	3	8	10
21	30	2	5	4	7	17	23	4	9
10	1	4	3	6	16	22	28	32	5

(a) Set up a frequency distribution of the number of trials required to learn the lesson, using the intervals 1 to 5, 6 to 10, 11 to 15, 16 to 20, 21 to 25, 26 to 30, and 31 to 35.

(b) Construct a frequency polygon from the frequency distribution of part a.

SECTION 1.B
AVERAGES

Consider a small business employing 10 persons whose weekly salaries are listed in Table 1.5. One question that may be of interest is, "What is the average weekly salary of all those on the payroll?"

The usual method of computing the "average" salary is to divide the total number of dollars paid out weekly by the total number of persons on the payroll. Let's do it. The total number of dollars paid out is 5000, as shown at the bottom of Table 1.5, while there are 10 persons on the payroll. This method of computing the average yields an average weekly salary of 5000/10 = 500 dollars. So far so good.

The next questions that arise are, "What does the number 500 signify? What information does it communicate about the data? Why did we compute the average salary by such a procedure? What, in fact, are we talking about when we mention the average salary?" Consider the following proposals for the meaning of the word "average":

1. The average salary is the middle salary: Half of those on the payroll earn at or less than the average salary, while the other half earn at or more than the average salary.

2. The average salary is the most common salary: More people earn that salary than any other single dollar amount.

TABLE 1.5
Weekly salaries

Person	Weekly Salary, dollars
A	400
B	200
C	100
D	200
E	2500
F	200
G	300
H	600
I	200
J	300
Total payroll = 5000	

~~While it would be difficult to think up another possible interpretation of the word "average," it is a fact that the average we have computed, namely 500 dollars, satisfies neither of the above criteria.~~ To see this, let's calculate the specific numbers which play the roles defined by the above descriptions of the word "average."

First, ~~the middle salary, which is technically referred to as the "median,"~~ can be determined by listing the data points from lowest to highest and finding a number located exactly in the middle. From Table 1.6, it is apparent that the median of this set of data is the number 250, for persons C, D, F, B, and I earn at or less than the 250-dollar level, while persons G, J, A, H, and E earn at or more than the 250-dollar level. ~~(If there is an odd number of data points, the median will be the exact middle number, while if there is an even number of data points, the median is usually taken to be the number halfway between the two middle~~

TABLE 1.6
Weekly salaries in numerical order

Person	Weekly Salary, dollars
C	100
D	200
F	200
B	200
I	200
G	300
J	300
A	400
H	600
E	2500

TABLE 1.7
Persons grouped by salary level

Salary Level, dollars	No. of Persons at That Salary Level
100	1 (C)
200	4 (D, F, B, I)
300	2 (G, J)
400	1 (A)
600	1 (H)
2500	1 (E)
Sum = 10	

numbers. In this case, the two middle numbers are 200 and 300, and so we consider 250 to be the median.) The average that we computed earlier, namely 500, is certainly not the median, since we now know that the median salary is 250 dollars. From Table 1.6, we can see further that 8 out of the 10 persons on the payroll are earning less than the so-called average salary of 500 dollars.[1]

Second, the most common salary, which is technically called the "mode," is determined by listing each possible salary level and then noting the number of persons receiving that salary. From Table 1.7, we see that the "modal" salary is 200 dollars because four persons have salaries at that level. This is far more than the number of persons at any other salary level. It is interesting to observe that, for this set of data at least, the modal salary of 200 dollars, the median salary of 250 dollars, and what we have been calling the average salary of 500 dollars are all different numbers.

What, then, is the meaning of the average which is calculated by summing all data points and dividing that sum by the number of data points involved? Well, the precise significance of that process is not easy to explain in a few words; and so we will postpone the details until the next section. The technical name of the average computed in this manner is the "mean." In particular, 500 dollars is the mean salary of all persons on the payroll of the small business under study. In Table 1.8, we summarize the various types of averages discussed in our attempt to determine the average salary of persons on the payroll listed in Table 1.5.

In view of the fact that the word "average" may mean several different things, and often different things to different people, how do we know which type of average is appropriate in any particular situation? As a rule, this decision is not based on mathematical considerations but instead on criteria existing within the applied field of study. For example, geographers talk about the mean annual temperature or mean annual rainfall in a particular geographical area.

---

[1] This observation has the somewhat entertaining sociological interpretation that nearly everybody is below average.

TABLE 1.8
"Averages" of weekly salaries

Type of Average	Numerical Value of Average
Mean	500
Median	250
Mode	200

Economists and sociologists are interested in the median family income of a certain community. Shoe and clothing outlets have to make sure they stock the modal sizes (for there may be relatively few individuals having exactly the mean or median sizes, and possibly no individuals at all with these sizes, as a comparison of Tables 1.5 and 1.8 will illustrate).

EXERCISES 1.B

1 The following data are the prices, in cents per pound, of imported English cheese in 12 metropolitan areas of the United States and Canada:

100	105
110	110
130	100
120	90
110	115
90	104

(a) Calculate the mean price.
(b) Determine the median price.
(c) Find the modal price.

2 The data listed below represent levels of impurity in each of 16 lots of tea leaves:

5	6
4	5
25	7
9	10
11	6
7	8
12	5
20	4

(a) Calculate the mean impurity level.
(b) Determine the median impurity level.
(c) Find the modal impurity level.

3　As part of a study aimed at finding out what a typical auto-body repair job costs, an insurance adjusters' organization collected the following data on repair estimates, in dollars, for 15 damaged cars:

220	200
460	380
630	590
300	280
120	130
540	500
730	650
930	

(a)　What is the mean repair estimate?
(b)　What is the median repair estimate?
(c)　Is there a modal repair estimate?

4　Use trial-and-error to make up a set of three data points whose mean is larger than its median.

5　Make up a set of three data points whose mean is smaller than its median.

6　Make up a set of five data points having the same mean and median but having a mode different from the mean and median.

7　Just as the median divides a set of data into two parts, a lower half and an upper half, we can define quartiles (which divide the data into four equal parts), deciles (10 equal parts), and percentiles (100 equal parts). For example, the 7th decile divides the lower 70% of the data from the upper 30%, while the 3rd quartile divides the lower 75% from the upper 25%. Using the data of Exercise 1.A.5, find

(a)　The median
(b)　The 1st quartile
(c)　The 3rd decile
(d)　The 57th percentile

8　From the data of Exercise 1.A.1, determine

(a)　The median
(b)　The 3rd quartile
(c)　The 9th decile
(d)　the 38th percentile

SECTION 1.C
THE MEAN

Despite the fact that the mean (calculated by summing the data points and then dividing the sum by how many of them there are) is the most commonly used measure of average, it is not clear exactly what information it conveys. In the present section, we will discuss the significance of the mean as a measure of average.

THE BALANCING PROPERTY OF THE MEAN

Let's look again at the data on weekly salaries appearing in Table 1.5. Imagine a seesaw in a children's playground marked off in units of measure. We place bowling balls at locations corresponding to the data points, and we obtain a graphical description of the information presented in Table 1.7. The picture of the seesaw appears in Fig. 1.3, with the mean value of 500 dollars clearly marked. In the picture, we see one ball at location 100, four at 200, two at 300, one at 400, one at 600, and one at 2500, in accordance with Table 1.7. What we will now show is that the seesaw will balance only if its fulcrum (balancing point) is placed at the mean value, namely, 500 in this situation.

One of the principles of the seesaw asserts that a 25-pound child will be able to balance a 50-pound child if the smaller child sits twice as far from the center as does the larger child. In a similar spirit, we see that four bowling balls at a distance 300 from the center will balance one bowling ball at a distance 1200 from the center. You could draw a diagram of this, or could actually test it out (using bricks instead of bowling balls) at a local playground. In the seesaw of Fig. 1.3, the balls at 400 and 600 balance each other around the center position of 500, while the one ball at 2500 balances the remaining seven balls.

It is in this sense, the sense of the balancing point of the data, that the mean can be considered a measure of average. When we are informed that the mean of a particular set of data is 500, we should visualize in our minds the data strung out along a seesaw which is balanced at 500.

Let's try to organize this balancing property into a useful technique for describing data. If we subtract the mean value of 500 from each of the data points, for example, subtract 500 from 300, we get a number which tells us the distance of the data point from the mean. It also tells us on which side of the mean the data point is located. Using as an example the data point 300, we calculate $300 - 500 = -200$. The negative sign indicates that 300 is to the left of the mean on the seesaw. The 200 indicates that 300 is a distance of 200 units away from the mean. Similarly, $600 - 500 = 100$ indicates that the point 600 is 100 units to the right of the mean.

We call the difference between a data point and the mean the "deviation of the data point from the mean." The deviation will be negative if the data point

FIGURE 1.3
Balancing property of the mean weekly salary.

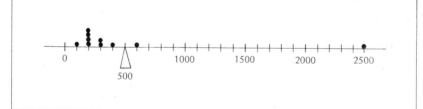

TABLE 1.9
Sum of deviations of weekly salary data

Person	Weekly Salary, dollars	Salary − Mean =	Deviation of Salary
A	400	400 − 500 =	−100
B	200	200 − 500 =	−300
C	100	100 − 500 =	−400
D	200	200 − 500 =	−300
E	2500	2500 − 500 =	2000
F	200	200 − 500 =	−300
G	300	300 − 500 =	−200
H	600	600 − 500 =	100
I	200	200 − 500 =	−300
J	300	300 − 500 =	−200
		Sum of deviations =	0

falls to the left of the mean (on the seesaw) and positive if the data point lies to the right of the mean. The balancing property means that the positive deviations exactly cancel the negative deviations, so that the sum of all deviations is zero. In Table 1.9, we show how the deviations sum to zero in the case of the weekly salary data of Table 1.5.

The deviations of the various data points from the mean are very important in themselves, for they can be used to measure whether the data points are all bunched together near the mean or whether some are dispersed far from the mean. We will deal with this question in detail in the next section. Here let us remark only that the computational procedure carried out in Table 1.9 can be used as a check on our calculation of the mean. If the sum of the deviations turns out to be something other than zero, we have probably made an arithmetic error in our calculation of the mean.

LABELING A SET OF DATA POINTS

In order to establish general procedures for analyzing a set of data, it is necessary to develop new techniques of labeling a set of data points. One convenient way of labeling a set of data points is to consider the data as a set of x's labeled as follows:

x_1 = first data point

x_2 = second data point

x_3 = third data point

....

x_k = kth data point   (pronounced "kayth")

....

x_n = nth data point   (pronounced "enth")

We will agree that x_n will be the last of our data points, and this means that we are dealing with n data points, so that

n = number of data points in our set

In the case of the weekly salary data of Table 1.5, we have 10 data points so that $n = 10$. We label the points of Table 1.5 as follows:

$x_1 = 400$ $x_6 = 200$

$x_2 = 200$ $x_7 = 300$

$x_3 = 100$ $x_8 = 600$

$x_4 = 200$ $x_9 = 200$

$x_5 = 2500$ $x_{10} = 300$

To avoid the necessity of writing the words "sum of the data points" over and over again, we should introduce some symbols for that expression. The symbol used in statistics to signify the word "sum" is the capital letter Σ (pronounced "sigma") of the Greek alphabet. To symbolize the sum of the data points, we write Σx, which is pronounced "the sum of the x's." The meaning of those symbols is therefore

$$\Sigma x = x_1 + x_2 + x_3 + \cdots + x_n$$

For the weekly salary data

$$\Sigma x = x_1 + x_2 + x_3 + x_4 + x_5 + x_6 + x_7 + x_8 + x_9 + x_{10}$$

$$= 400 + 200 + 100 + 200 + 2500 + 200 + 300 + 600 + 200 + 300$$

$$= 5000$$

Using statistical "shorthand," we can write a compact expression for the mean. In situations where the data represent a complete "population" (as do the weekly salaries of *all* 10 persons employed in the small business under study) rather than a statistical sample,[1] we use the lowercase Greek letter μ (pronounced "mew") as a symbol for the mean. The expression

$$\text{The mean} = \frac{\text{the sum of all data points}}{\text{the number of data points}}$$

can then be translated into our new shorthand as

$$\mu = \frac{\Sigma x}{n}$$

[1] We will study statistical samples in Chap. 3.

TABLE 1.10
Sum of deviations from the mean

	Data Points x	Deviations $x - \mu$
	x_1	$x_1 - \mu$
	x_2	$x_2 - \mu$
	x_3	$x_3 - \mu$
	$\vdots$	$\vdots$
	x_n	$x_n - \mu$
Sums	Σx	0

With the shorthand formula, we can express the mean of the weekly salary data as

$$\mu = \frac{\Sigma x}{n} = \frac{5000}{10} = 500$$

In terms of our system of labeling data points, we can construct a table, analogous to Table 1.9, which illustrates the balancing property by pointing out that the deviations always have a sum of zero. This symbolic calculation appears in Table 1.10. We write the deviation from the mean (found by subtracting the mean from the data point) as $x - \mu$.

Because of the balancing property, it turns out that $\Sigma(x - \mu) = 0$ for any possible set of data points; that is, the sum of the deviations is zero. (Try a few sets for yourself!)

EXERCISES 1.C

1 Draw a seesaw representation of the cheese price data of Exercise 1.B.1, and show that the deviations from the mean really do sum to zero.

2 Construct a seesaw representation of the impurity level data of Exercise 1.B.2, and make the calculations which verify that the deviations from the mean have a sum of zero.

3 Put the auto repair estimates of Exercise 1.B.3 on a seesaw, and show that the deviations from the mean sum to zero.

4 Make up a set of any nine numbers you can think of, and verify that the deviations from the mean really sum to zero.

5 Using the data of Exercise 1.B.1,
   (a) Show that the deviations from the *median do not* sum to zero.
   (b) Show that the deviations from the *mode do not* sum to zero.

6 Using the data of Exercise 1.B.2,
   (a) Show that the deviations from the *median do not* sum to zero.

(b) Show that the deviations from the *mode do not* sum to zero.

7 Using the data of Exercise 1.B.3, show that the deviations from the *median do not* sum to zero.

8 Using the data of Exercise 1.B.6, show that the deviations from the *median do* sum to zero. Can you explain why this occurs?

9 If $x_1 = 5$, $x_2 = 8$, $x_3 = 7$, and $x_4 = 19$, calculate Σx.

10 If

$$x_1 = 17 \qquad x_5 = -8$$

$$x_2 = 9 \qquad x_6 = -9$$

$$x_3 = -6 \qquad x_7 = -5$$

$$x_4 = 14 \qquad x_8 = 3$$

calculate Σx.

SECTION 1.D
THE STANDARD DEVIATION

If a geographer compares two regions of a continent on the same day in regard to their recorded temperatures at various times during the day, she might come up with the data presented in the top part of Table 1.11. A check of the bottom part of Table 1.11 shows that, no matter what you consider the appropriate measure of average, each set of data has an average of 50°F. It would therefore be important to know exactly how descriptive of a set of data an average is. In particular, when we are informed that the average of a certain set of data is 50, how are we to interpret that information? How are we to visualize or reconstruct in our mind the original set of data from the knowledge that the average is 50?

As the pictorial representation shown in Fig. 1.4 demonstrates, two sets of

TABLE 1.11
Comparison of regional temperatures

Hour	East Coast Temperatures, °F	West Coast Temperatures, °F
3 A.M.	40	10
6 A.M.	50	50
9 A.M.	50	50
12 P.M.	60	80
3 P.M.	50	90
6 P.M.	50	20
Mean	50	50
Median	50	50
Mode	50	50

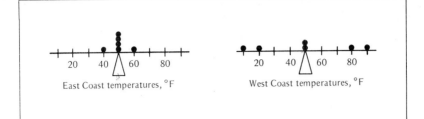

FIGURE 1.4
"Seesaw" representations of temperature data.

data do not necessarily exhibit the same overall characteristics just because they both have an average of 50°F. How then do the two sets of temperature data differ? As Fig. 1.4 shows, they differ primarily by how much the data points vary from the common average of 50°F. The East Coast temperatures are heavily concentrated near their average of 50°F, while the West Coast temperatures vary considerably from their average of 50°F. In particular, the East Coast temperatures remain between 40 and 60°F throughout the day, while West Coast temperatures range from a low of 10 to a high of 90°F. Knowledge of the averages alone gives no information about the variation.

It would be very useful, therefore, to have a numerical measure of variation of the individual points within a set of data. If we report the variation each time we report the average, we will communicate a more accurate description of the general pattern of the data.

MEASURES OF VARIATION

Consider the following proposal for a measure of variation of a set of data: We can look at the average deviation of points from their mean, technically referred to as the "mean absolute deviation." To compute it involves the use of a table of deviations, like those in Table 1.9. The preliminary computations leading to the mean absolute deviation of each set of temperature data can be found in Table 1.12. In the portion of Table 1.12 dealing with East Coast temperatures, the column headed x lists the actual temperature data, while the column headed $x - \mu$ lists the deviation of each data point from the mean value of 50. In view of the balancing property of the mean as illustrated in Table 1.10, we know that the column headed $x - \mu$ must sum to zero. To measure variation we need the absolute magnitude (technically called the "absolute value") of the deviations which are listed in the column headed $|x - \mu|$.[1] The mean absolute deviation is

[1]By the "absolute value" of a number, we mean the magnitude of the number, ignoring the + or − sign. For example, both $|5| = 5$ and $|-5| = 5$. The upshot of this is that, whether a number is positive or negative (or zero), its absolute value cannot be negative.

TABLE 1.12
Computations leading to the mean absolute deviations of temperature data

	East Coast Temperatures x			West Coast Temperatures y		
	x	$x - \mu$	$\|x - \mu\|$	y	$y - \mu$	$\|y - \mu\|$
	40	-10	10	10	-40	40
	50	0	0	50	0	0
	50	0	0	50	0	0
	60	10	10	80	30	30
	50	0	0	90	40	40
	50	0	0	20	-30	30
Sums	300	0	20	300	0	140

$n = 6$
$\mu = \frac{300}{6} = 50$

$n = 6$
$\mu = \frac{300}{6} = 50$

then the sum of the $|x - \mu|$ column divided by the number of data points. A set of data with its data points far from their mean will have a larger mean absolute deviation than a set having its points closer to their mean.

We abbreviate the mean absolute deviation by the symbol MAD. With our statistical symbols, we can write

$$\text{MAD} = \frac{\text{sum of absolute deviations}}{\text{number of data points}} = \frac{\Sigma|x - \mu|}{n}$$

The East Coast temperatures have MAD = 20/6 = 3.33, using the fact in Table 1.12 that the sum of the absolute deviations (the column headed by $|x - \mu|$) is 20. On the other hand, the West Coast temperatures have MAD = 140/6 = 23.33. This shows that hourly West Coast temperatures are more variable than hourly East Coast temperatures.

Now that we have thoroughly discussed the use and computation of the mean absolute deviation as a possible measure of the variation of a set of data points, it is the author's sad duty to report that this measure is not widely used, except in very special cases involving extremely large variations of the data. It has been preempted in importance by a second measure of variation, technically referred to as the "standard deviation." The standard deviation is somewhat more complicated to calculate than the MAD, and also generally more useful and more efficient.

The standard deviation is the square root of the mean of the squared deviations. (In some applied contexts, it is referred to as the "root-mean-square deviation" or the "rms deviation.") We usually denote the standard deviation of an entire population (as opposed to a statistical sample) by the Greek letter σ (lowercase "sigma"), and in mathematical symbolism, we express it as follows:

$$\sigma = \sqrt{\frac{\Sigma(x - \mu)^2}{n}}$$

TABLE 1.13
Computations leading to the standard deviations of temperature data

	East Coast Temperatures x			West Coast Temperatures y		
	x	$x - \mu$	$(x - \mu)^2$	y	$y - \mu$	$(y - \mu)^2$
	40	−10	100	10	−40	1600
	50	0	0	50	0	0
	50	0	0	50	0	0
	60	10	100	80	30	900
	50	0	0	90	40	1600
	50	0	0	20	−30	900
Sums	300	0	200	300	0	5000

$$n = 6 \qquad\qquad\qquad\qquad n = 6$$
$$\mu = \tfrac{300}{6} = 50 \qquad\qquad\quad \mu = \tfrac{300}{6} = 50$$

To calculate σ in accordance with the above formula, we proceed as follows:

1. Calculate the deviations $x - \mu$ as in Table 1.12.
2. Square each deviation; i.e., multiply it by itself.
3. Sum the squared deviations.
4. Divide the sum by n, the number of data points.
5. Take the square root[1] of the resulting number.

The standard deviation is the most widely used measure of variation because it conveys the most precise and useful information about variation of the data from the mean. The sort of information it conveys will be discussed in detail in the next two sections and in portions of the next chapter. Here we concentrate on how to calculate it.

The most direct method of calculating the standard deviation is by means of a table of preliminary computations analogous to Table 1.12. In accordance with the five-step procedure discussed above, we would record the square of each deviation rather than its absolute value. A deviation of −10 would have a square of $(-10) \times (-10) = 100$, if we recall that the product of two negative numbers is a positive number. Similarly, a deviation of 40 has a square of $40 \times 40 = 1600$. We construct Table 1.13 with these principles in mind. From Table 1.13, we see that the sum of the squared deviations of the East Coast temperatures is

$$\Sigma(x - \mu)^2 = 100 + 0 + 0 + 100 + 0 + 0 = 200$$

by adding up the column labeled $(x - \mu)^2$. The sum of the squared deviations of West Coast temperatures is, by adding up the column labeled $(y - \mu)^2$,

---

[1] The square root may be explained as follows: The square of a number, say 3, is that number multiplied by itself; namely, $3 \times 3 = 9$. The square root is the reverse of this; that is, $\sqrt{9} = 3$ because $3 \times 3 = 9$. In short, for positive numbers a and b, $\sqrt{a} = b$ if and only if $b^2 = a$.

$$\Sigma(y - \mu)^2 = 1600 + 0 + 0 + 900 + 1600 + 900 = 5000$$

It follows that, for East Coast temperatures, the standard deviation is

$$\sigma = \sqrt{\frac{\Sigma(y - \mu)^2}{n}} = \sqrt{\frac{200}{6}} = \sqrt{33.3} = 5.77$$

West Coast temperatures have standard deviation

$$\sigma = \sqrt{\frac{\Sigma(y - \mu)^2}{n}} = \sqrt{\frac{5000}{6}} = \sqrt{833} = 28.86$$

As is clear from Fig. 1.4, the West Coast temperatures are more variable than the East Coast's. Correspondingly, the West Coast temperatures' standard deviation of 28.86 exceeds by a considerable margin the East Coast's value of 5.77.

SQUARE ROOTS

Before proceeding further in our study of the standard deviation, let's pause a moment to talk about the square roots involved in the calculation. There are basically three ways of finding the square root of a number: (1) using an electronic calculator which has a square-root button; (2) carrying out the paper-and-pencil calculation by hand using the long-division-type method sometimes taught in elementary algebra courses; and (3) using the square-root tables in the back of this textbook. Method 1 is obviously the most efficient and most accurate method. It is the one used by those persons who handle statistics in their daily work. However, not every calculator has a square-root button, and so method 1 will not be practical for everyone who uses this text. The hand calculation of method 2 is somewhat more complicated and time-consuming than ordinary long division. Therefore, we also consider this method impractical, although it is a fascinating experience to work out the square root by yourself. If you are interested in the details of the hand calculation method, you are invited to ask your local mathematics expert, namely your instructor, for an explanation of it. This brings us to method 3. Using Table A.1 of the Appendix, we can work out the square root of any number having three significant digits. Because Table A.1 contains only the numbers 1.00 through 9.99 explicitly, it is necessary to express the number whose square root we want in terms of one of those numbers. To illustrate this procedure, we will work out the details of finding $\sqrt{33.3}$ and $\sqrt{833}$. We have

$$\sqrt{33.3} = \sqrt{3.33 \times 10} = 5.77$$

looking up $N = 3.33$ and using the third column of the table, the column headed $\sqrt{10N} = \sqrt{N \times 10}$. Next,

$$\sqrt{833} = \sqrt{8.33 \times 100} = \sqrt{8.33} \times \sqrt{100}$$

$$= 2.886 \times 10 = 28.86$$

looking up $N = 8.33$, finding $\sqrt{N} = 2.886$, and realizing that you probably have already memorized the fact that $\sqrt{100} = 10$. Additional examples throughout this text will continue to illustrate the procedure of finding square roots using Table A.1.

SHORTCUT METHOD

In some situations in statistics, it is more efficient to calculate the standard deviation by an alternative procedure. The alternative procedure, euphemistically called "the shortcut method" of computing the standard deviation, does in fact simplify the computation when the mean μ does not come out "even" (namely, as a whole number). For example, consider the data of Table 1.14 showing the weekly sales volume of a small auto-parts outlet over a 7-week period. The mean sales volume is

$$\mu = \frac{\Sigma x}{n} = \frac{45}{7} = 6.43$$

which does not come out even. In Table 1.15, we illustrate the procedures for calculating the standard deviation by both the direct and the shortcut methods. You should observe that the shortcut method does indeed save some steps in this case. The shortcut formula does not require the mean to be computed first, but rather it proceeds directly to the standard deviation. The formula involves only the sum of the data points and the sum of their squares:

$$\sigma = \sqrt{\frac{n \, \Sigma \, x^2 - (\Sigma x)^2}{n^2}}$$

TABLE 1.14
Weekly sales of auto parts

Week	Sales Volume, thousands of dollars
#1	7
#2	5
#3	7
#4	8
#5	4
#6	9
#7	5
	Sum = 45

TABLE 1.15
Calculation of the standard deviation of auto parts sales data, direct method versus shortcut method

	Direct Method			Shortcut Method	
x	$x - \mu$	$(x - \mu)^2$		x	x^2
7	0.57	0.3249		7	49
5	−1.43	2.0449		5	25
7	0.57	0.3249		7	49
8	1.57	2.4649		8	64
4	−2.43	5.9049		4	16
9	2.57	6.6049		9	81
5	−1.43	2.0449		5	25
Sums 45	−0.01	19.7143		45	309

$n = 7$ $n = 7$

$\mu = \frac{45}{7} = 6.43$

Both the direct and the shortcut formulas for the standard deviation give exactly the same answer in every case, except for possible differences due to rounding off decimal places. Notice that some round-off error has already occurred in the computations preparatory to the direct method. Such an error is indicated by the fact that the deviations sum to −.01 instead of exactly 0 as the balancing property requires. The error cannot be avoided because, as a practical matter, we have to round off μ to 6.43 from its more precise value of 6.428571429. Therefore we might, in some cases, wind up with a slightly inaccurate value of σ as well. The shortcut method, however, maintains 100% accuracy in the computation table and is therefore the preferred method of computing σ when μ does not come out "even." Furthermore, the shortcut formula is easier to use when computing the standard deviation using a calculator, for basically all that is needed are the sum of the data points and the sum of their squares. We now calculate σ both ways from the information in Table 1.15:

Direct method: $\sigma = \sqrt{\dfrac{\Sigma(x - \mu)^2}{n}} = \sqrt{\dfrac{19.7143}{7}} = \sqrt{2.82} = 1.68$

Shortcut method: $\sigma = \sqrt{\dfrac{n\,\Sigma\,x^2 - (\Sigma x)^2}{n^2}} = \sqrt{\dfrac{7(309) - (45)^2}{7^2}}$

$= \sqrt{\dfrac{2163 - 2025}{49}} = \sqrt{\dfrac{138}{49}} = \sqrt{2.82} = 1.68$

You should observe that both methods yield the same value of σ, namely 1.68. It is also important to observe that Σx^2 and $(\Sigma x)^2$ are not one and the same. To get Σx^2, we sum the squares, while to get $(\Sigma x)^2$, we square the sum. Here $\Sigma x^2 = 309$, and $(\Sigma x)^2 = 2025$. They're quite different.

EXERCISES 1.D

1 From the data on prices of English cheese appearing in Exercise 1.B.1, calculate
   - (a) The standard deviation of the prices
   - (b) The mean absolute deviation
   - (c) The standard deviation by using the shortcut formula
2 Using the data points given in Exercise 1.B.2, find
   - (a) The standard deviation of the impurity levels by computations based on the direct formula
   - (b) The standard deviation by computations based on the shortcut formula
   - (c) The mean absolute deviation of the impurity levels
3 On the basis of the data of Exercise 1.B.3, calculate
   - (a) The standard deviation of the repair estimates using the direct formula for the standard deviation
   - (b) The standard deviation using the shortcut formula
   - (c) The mean absolute deviation

SECTION 1.E
CHEBYSHEV'S THEOREM

When we want to communicate information about the data, we ideally want our listener to be able to reconstruct the frequency polygon of the data after we finish talking. If we present the information, but the listener does not get a good idea of the nature of the data, then we have wasted our time making the presentation. Suppose you hear that Fosbert Realty takes a mean time of 14 days to sell the houses it lists, with a standard deviation of 4 days. How well do you understand the data? Do you understand the data well enough to know, for example, how many of Fosbert's houses sell within 25 days or how many stay on the market longer than 30 days? Chebyshev's theorem provides useful answers to both of these questions.

Chebyshev's theorem may be stated as follows:

If a set of data points has mean μ and standard deviation σ, then the proportion p of data lying farther from the mean than k standard deviations cannot exceed $1/k^2$.

What does this statement mean? Figure 1.5 illustrates what Chebyshev's theorem is saying. Forming a length of k standard deviations, namely $k\sigma$, on both sides of the mean μ, Chebyshev's theorem says that not more than a fraction $1/k^2$ of the data points can possibly fall outside this interval. This implies two equivalent assertions:

1 Not more than a proportion $1/k^2$ of the data points can have numerical values smaller than $\mu - k\sigma$ or greater than $\mu + k\sigma$.

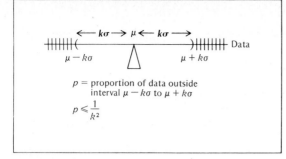

FIGURE 1.5
Chebyshev's theorem.

2 At least a proportion $1 - 1/k^2$ of the data points must have values between $\mu - k\sigma$ and $\mu + k\sigma$.

The second assertion is a logical consequence of the first because if at most a fraction $1/k^2$ of the data falls outside the interval, then at least the remainder $1 - 1/k^2$ must fall inside. For example, considering an interval of 3 standard deviations about the mean, we see that $k = 3$, and we know that no more than $1/k^2 = \frac{1}{9}$ of the data can fall outside the interval. This automatically implies that at least $1 - 1/k^2 = 1 - \frac{1}{9} = \frac{8}{9}$ must be falling inside. This situation is illustrated graphically in Fig. 1.6.

Let's take another look at the questions involving Fosbert Realty, where the relevant data have mean $\mu = 14$ and standard deviation $\sigma = 4$. The question of how many houses remain on the market longer than 30 days can be transformed into the picture of Fig. 1.7. We are interested in the proportion of data points falling above 30. But 30 lies at a distance of $30 - 14 = 16$ from the mean $\mu = 14$, and so we are interested in the proportion of data points falling farther from the mean than a distance of 16. Because $\sigma = 4$, the distance 16 represents a distance of $k = 16/\sigma = \frac{16}{4} = 4$ standard deviations. By Chebyshev's theorem, the proportion of data points farther from the mean than $k = 4$ standard deviations

FIGURE 1.6
Chebyshev's theorem with $k=3$.

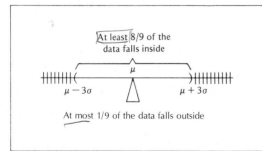

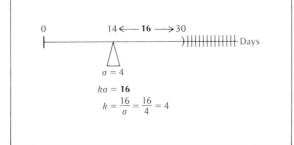

FIGURE 1.7
Fosbert Realty data: Proportion of houses on market longer than 30 days.

cannot exceed $1/k^2 = 1/4^2 = \frac{1}{16} = .0625 = 6.25\%$. Therefore, we know that no more than 6.25% of Fosbert Realty's houses remain unsold after 30 days on the listing.

The second question involving the Fosbert Realty data asks what proportion of Fosbert's houses sell within 25 days. In Fig. 1.8 we illustrate this question. The mean $\mu = 14$, and so the relevant number 25 lies at a distance $25 - 14 = 11$ from the mean. Since $\sigma = 4$, the distance 11 can be expressed as $k = \frac{11}{4} = 2.75$ standard deviations. Chebyshev's theorem then asserts that no more than $1/k^2 = 1/(2.75)^2 = 1/7.5625 = .132 = 13.2\%$ of the data points can lie farther from the mean than 2.75 standard deviations. This means that no more than 13.2% of the data points can fall outside the interval 3 to 25. In particular, no more than 13.2% of Fosbert Realty's houses are sold on days outside the interval 3 to 25 days after listing. Therefore, at least 86.8% (= 100% minus 13.2%) of the houses are sold between 3 and 25 days after listing. We can therefore be sure that at least 86.8% are sold within 25 days of listing.

It is possible that more than 86.8% are sold within 25 days because the 86.8% figure does not include any houses sold within the first 3 days. Houses in

FIGURE 1.8
Fosbert Realty data: Proportion of houses selling within 25 days.

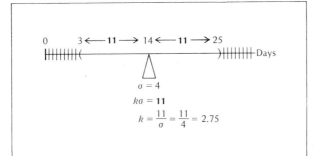

TABLE 1.16
Information given by Chebyshev's theorem

Distance from Mean to End of Interval in Standard Deviations k	Maximum Possible Percentage of Data Outside Interval $1/k^2$	At Least This Much Data Falls Within Interval $1 - 1/k^2$
1	100.00%	0.00%
2	25.00%	75.00%
3	11.11%	88.89%
4	6.25%	93.75%
5	4.00%	96.00%
6	2.78%	97.22%
7	2.04%	97.96%
8	1.56%	98.44%
9	1.23%	98.77%
10	1.00%	99.00%

the latter category are included among the 13.2% lying outside the interval. Unfortunately, Chebyshev's theorem does not provide any breakdown of the percentage falling above 25 and below 3, but only percentages falling outside and inside various intervals.

We can summarize the sort of information available from Chebyshev's theorem in a table such as Table 1.16, which is based on Fig. 1.5. You should observe from Table 1.16 that, no matter what the set of data, no more than 1% of the data points can fall farther from the mean than a distance of 10 standard deviations, and no more than 4% of the data points can fall farther away than 5 standard deviations. It automatically follows that at least 96% of the data must fall within 5 standard deviations, and at least 99% within 10 standard deviations. Analogous statements can be made for any number k of standard deviations, regardless of the shape of the frequency polygon of the data.

Example 1.1 Job Aptitude Test   Eighty new jobs are opening up at an airplane manufacturing plant, but 1100 applicants show up for the 80 positions. To select the best 80 from among the applicants, the prospective employer gives a combination physical and written aptitude test which covers mechanical skill, manual dexterity, and mathematical ability. The mean grade on this test turns out to be 175, and the scores have standard deviation 10. Can a person who scored 215 count on getting one of the jobs?

SOLUTION   From the way the question is formulated, what we need to find out is the number of persons scoring above 215. This information would be useful because, if fewer than 80 applicants scored above 215, then a person who scored 215 would be among the top 80 and would therefore get one of the open positions. From Fig. 1.9, we see that the distance $k\sigma$ from the mean of 175 to the end

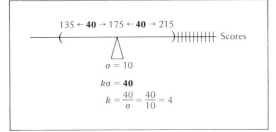

FIGURE 1.9
Aptitude test scores.

of the interval at 215 is $215 - 175 = 40$. However, $k\sigma = 40$ means that $k = 40/\sigma = \frac{40}{10} = 4$ and $1/k^2 = 1/4^2 = \frac{1}{16} = .0625 = 6.25\%$. According to Chebyshev's theorem, this means that no more than 6.25% of the applicants could have obtained scores differing from 175 by more than 40 points. That is, if we look at all those who scored below 135 and all those who scored above 215, then both those will account for no more than 6.25% of all applicants. In particular, the most we can say is that no more than 6.25% could have scored above 215, in view of the fact that Chebyshev's theorem does not allow us to say how much of the 6.25% falls below 135 and how much above 215. Therefore, because 6.25% of 1100 equals .0625 times $1100 = 68.75$, we can be sure that no more than 68 of the applicants could have scored higher than 215. (Even though 68.75 is rounded to 69, there cannot be 69 applicants scoring above 215, because no more than 68.75 could have done so.) Since there are 80 positions, a person scoring 215 indeed gets one of the jobs, for no more than 68 people could have scored higher.

Example 1.2 Contents of Cereal Boxes The law regulating producers of 9-ounce boxes of breakfast cereal asserts that no more than one-half of 1% of the output of any brand of cereal (in boxes labeled as containing 9 ounces) are allowed to contain less than 8.8 ounces. One company markets a 9-ounce box of Corn Puffies which is packed by machine in such a way that the boxes have mean weight 9.0 ounces with a standard deviation of .01 ounce. Is the company in violation of the law?

SOLUTION   To test whether the law is being obeyed, it is necessary to find out what proportion of 9-ounce boxes of Corn Puffies contain less than 8.8 ounces. If this proportion turns out to be below .005 (one-half of 1%), then the company is conducting a legal operation. If not, some further investigation might be required. In Fig. 1.10, we are interested in the proportion of data falling in the marked region, i.e., the proportion of data less than 8.8. Here the distance $k\sigma$ from the mean of 9.0 to the end of the interval at 8.8 is .2. But $k\sigma = .2$ means that $k = .2/\sigma = .2/.01 = 20$ and $1/k^2 = 1/(20)^2 = \frac{1}{400} = .0025$. Therefore by Chebyshev's theorem, no more than .0025 (one-quarter of 1%) of the Corn Puffies boxes will contain less than 8.8 ounces. Since a proportion up to .005 is allowed, the Corn

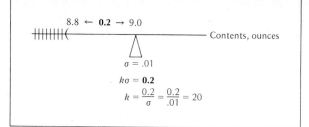

FIGURE 1.10
Cereal box contents (in ounces).

Puffies company is well within legal standards because its proportion of under-filled boxes will not exceed .0025.

EXERCISES 1.E

1 Even among cars produced on the same assembly line, rates of emission of environmental pollutants differ. In particular, the revolutionary new Peccarie (which gets 52 miles/gallon at freeway speeds) emits an average of 1.2 milligrams of pollutants per mile with a standard deviation of .04 milligram. New EPA regulations prohibit the sale of the entire production run of an automobile brand if more than 2% of the cars in the run emit pollutants in excess of 1.5 milligrams/mile. Can the Peccarie be legally sold in the United States?

2 State guidelines on the number of units taken by college students during their undergraduate careers will soon specify, in the interest of reducing costs, that no student ought to be permitted to exceed 220 quarter units of work or their equivalent (186 units are required for graduation). If graduating students at Bolsa Chica State average 203 units with a standard deviation of 4 units, at least what proportion of students already satisfy the new guidelines?

3 An electrical firm manufactures a 100-watt light bulb which, according to specifications printed on the package, has a mean life of 800 hours with a standard deviation of 40 hours. At most, what percentage of the bulbs fail to last even 700 hours?

4 Since the 55-mile/hour speed limit on freeways has been in effect, observers estimate that vehicles on the freeways travel with mean speed 55 miles/hour and standard deviation 2 miles/hour. If these estimates are accurate, at most, what percentage of the vehicles travel at speeds in excess of 65 miles/hour?

5 The 435 congressional districts in the United States have a mean population of 480,000 with a standard deviation of 30,000. At most, how many districts contain more than 600,000 residents?

6 Mean annual rainfall in a certain geographical region is 52.75 inches with a standard deviation of 6.50 inches. At most, what percentage of years have rainfall exceeding 70 inches?

7   A psychologist believes that it takes an average of 70 hours with a standard deviation of 10 hours to train a monkey to distinguish between the colors red, yellow, and blue. If his assumption is correct, would it be reasonable that out of a group of 360 monkeys, there were 80 monkeys able to learn the distinction in less than 40 hours of training?

8   The stock exchange is open 250 days each year. In a study of the fluctuation of the closing price of the stock of MGI, Inc., a computer readout indicated that the mean closing price for last year was 22.375 with a standard deviation of 5.000. At most, how many days did MGI stock close below 10?

9   After studying the incomes of families in a large urban area, a sociologist concludes that the mean income is 9000 dollars with a standard deviation of 2000 dollars. At most, what proportion of families have incomes higher than 25,000 dollars?

SECTION 1.F
STANDARD SCORES (z SCORES)

A certain graduate school requires all prospective students to take either the Graduate Record Examination (GRE) or the Montana Reasoning Test (MRT) as part of the application procedure. Unfortunately for comparison purposes, the GRE and the MRT are graded on different scales. The totality of GRE scores have mean 500 with a standard deviation of 100, while the MRT averages 60 with a standard deviation of 20. Table 1.17 shows the test scores submitted by five applicants, three of whom have taken the GRE, while the other two have taken the MRT. How is the school to compare and rank the scores, some of which are based on one type of scale and the rest on another? For example, how do we know which is the better score—a 700 on the GRE or a 110 on the MRT?

The way out of the problem is to adjust all the scores so that they will all be on the same scale. Then it will be easy to see which scores are higher than others, relatively speaking. Such an adjustment procedure is called "standardization" of scores, and the resulting adjusted scores are called "standard scores" or "z scores." To distinguish between the z scores and the original scores, the original ones are often referred to as "raw scores."

Standard scores can be best described as follows:

TABLE 1.17
Test scores submitted by applicants

Applicant	Test Taken	Score
A	GRE	700
B	GRE	450
C	MRT	70
D	GRE	600
E	MRT	110

If x is a data point of a set of data having mean μ and standard deviation σ, then the standard score (or z score) of the data point x is the number

$$z = \frac{x - \mu}{\sigma} = \frac{\text{deviation}}{\text{standard deviation}}$$

Notice that, to find the z score of a data point x, we subtract the mean of its group from x and divide the resulting difference by the standard deviation of its group.

Using the formula for z scores, let's find the z score for applicant A of Table 1.17. Applicant A scored 700 on the GRE, an exam which has mean 500 and standard deviation 100. Therefore her z score is

$$z_A = \frac{700 - 500}{100} = \frac{200}{100} = 2$$

On the other hand, let's take a look at applicant C. This individual scored 70 on the MRT, a test on which the mean score is 60 with a standard deviation of 20. Therefore, applicant C has z score

$$z_C = \frac{70 - 60}{20} = \frac{10}{20} = .5$$

The formula $z = (x - \mu)/\sigma$ really specifies the location of the point x in terms of the number of standard deviations it lies above the mean of its group. Therefore, applicant A scored 2 standard deviations above the mean of her group, while applicant C scored only .5 (one-half) standard deviation above the mean of his group. Relatively speaking, then, applicant A did better than applicant C.

To compute all the z scores, we set up a table of calculations like that of Table 1.18, recalling that GRE scores have a nationwide mean 500 and standard deviation 100, while MRT scores have a nationwide mean 60 and standard deviation 20. Having standardized all the scores, we can now easily rank all the

TABLE 1.18
Calculation of standard scores (z scores)

Applicant	Test	Raw Score x	Mean of Group, μ	Standard Deviation σ	$x - \mu$	z Score $\dfrac{x - \mu}{\sigma}$
A	GRE	700	500	100	200	2.0
B	GRE	450	500	100	−50	−.5
C	MRT	70	60	20	10	.5
D	GRE	600	500	100	100	1.0
E	MRT	110	60	20	50	2.5

TABLE 1.19
Applicants ranked by z scores

Rank	Applicant	z Score
1	E	2.5
2	A	2.0
3	D	1.0
4	C	.5
5	B	−.5

applicants. Applicant E had the highest score, 2.5 standard deviations above the mean of the MRT group, while applicant B got the lowest score, .5 standard deviation *below* the GRE group. The ranking of the applicants appears in Table 1.19.

Standardization of a set of data does the following thing: It transforms the original (raw) data into a set of numbers (called "z scores") which have mean 0 and standard deviation 1. In fact, no matter what the original values of μ and σ were, the z scores of a single set of data *always* have mean 0 and standard deviation 1. Once two or more sets of data are standardized, they can be graphed on the same scale and compared. This graphing process is illustrated in Fig. 1.11.

EXERCISES 1.F

1   Six applicants are applying for a single opening in medical technology at a research clinic working on skin cancer. Each applicant is given a 6-hour test

FIGURE 1.11
The process of standardization.

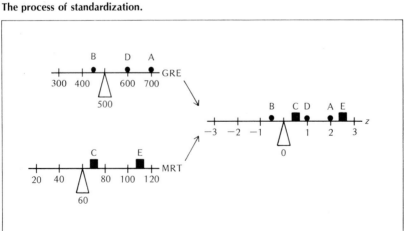

divided into three parts: ability to concentrate, manual dexterity, and knowledge of chemistry. Their grades on each part are as follows:

Applicant	Ability to Concentrate	Manual Dexterity	Knowledge of Chemistry	
A	10	9	5	24
B	9	9	6	24
C	5	2	0	7
D	5	9	10	24
E	1	0	6	7
F	0	1	9	10

In order to make fair comparisons, the grades are first standardized within each competency area, and the applicant with the highest total of z scores gets the job.

(a) Standardize the test scores of each applicant on the ability-to-concentrate section.

(b) Standardize all the manual dexterity scores.

(c) Standardize the knowledge-of-chemistry scores.

(d) Which applicant gets the job?

2 One college student took three exams on the same day. Her scores, as well as the mean and standard deviation of all the scores of her group, are presented in the following table:

Subject Matter	Student's Score	Group Mean	Group Standard Deviation
Economics	72	70.8	.6
Mathematics	61	51.0	6.0
Psychology	50	30.0	12.0

On which exam did the student do best in comparison with the rest of her group?

3 Four applicants for flight training, two from Portland, Oregon, and two from Tampa, Florida, have submitted their scores on different flight aptitude tests. The two from Oregon had taken the Great Open Skies Flight Aptitude Test (GOSFAT), while the pair from Florida had taken the Gulf and Southern Flight Aptitude Test (GASFAT). Scores on the GOSFAT have mean 50 with standard deviation 10, while those on the GASFAT have mean 120 and standard deviation 15. The applicants and their scores are listed below:

Applicant	Test Taken	Score
# 1	GOSFAT	60
# 2	GOSFAT	45
# 3	GASFAT	129
# 4	GASFAT	115

Standardize each score and rank the applicants.

4   Make up a set of five data points and standardize them. Then verify by calculation that the z scores have mean 0 and standard deviation 1.

SUMMARY AND DISCUSSION

Our objectives in Chap. 1 have been to introduce the role of statistical analysis in solving problems of applied value and to develop some methods of organizing and displaying statistical data. We have shown how a mass of statistical data, originally a mere collection of numbers, can be organized in tabular or graphical form to yield a comprehensive picture of the information contained within the data. The comprehensive picture of a single set of data is revealed through histograms and frequency polygons, with each type of illustration presenting a different perspective to the viewer. We further developed numerical methods of describing a set of data, mentioning three types of averages, the mean, the median, and the mode, and two measures of dispersion, the standard deviation and the mean absolute deviation, carefully noting the distinctions between these measures. Working with Chebyshev's theorem, we found that the mean and the standard deviation are useful for the measures of average and dispersion because when we apply Chebyshev's theorem to the mean and standard deviation of a set of data points, we can often answer very specific questions about an applied situation. Finally, we have discussed standardization of sets of data, with the aim of being able to compare widely differing sets of data by placing all of them on the same scale. The calculation of standard scores, or z scores, illustrated another important use of the mean and standard deviation. The material of Chap. 1 provides the framework for all the discussions of data in the remainder of the text.

BIBLIOGRAPHY

Statistical Sampling
Cochran, W. A.: "Sampling Techniques," 2d ed., Wiley, New York, 1963.
Deming, W. E.: "Some Theory of Sampling," Wiley, New York, 1950.
Raj, D.: "The Design of Sample Surveys," McGraw-Hill, New York, 1972.

Readings in the Applied Use of Statistics
Haber, A., et al. (eds.): "Readings in Statistics," Addison-Wesley, Reading, Mass., 1970.
Liebermann, B.: "Contemporary Problems in Statistics," Oxford University Press, New York, 1971.
Mansfield, E. (ed.): "Elementary Statistics for Business and Economics: Selected Readings," Norton, New York, 1970.
Tanur, J. M., et al. (eds.): "Statistics: A Guide to the Unknown," Holden-Day, San Francisco, 1972.

Wrigley, E. A. (ed.): "Nineteenth-Century Society: Essays in the Use of Quantitative Methods for the Study of Social Data," Cambridge University Press, Cambridge, England, 1972.

Nontechnical Discussions of Statistics
Bartholemew, D. J., and E. E. Bassett: "Let's Look at the Figures," Penguin Books, Baltimore, 1971.
Campbell, S. K.: "Flaws and Fallacies in Statistical Thinking," Prentice-Hall, Englewood Cliffs, N.J., 1974.
Huff, D.: "How to Lie with Statistics," Norton, New York, 1954.
Reichmann, W. J.: "Use and Abuse of Statistics," Oxford University Press, New York, 1961.
Von Mises, R.: "Probability, Statistics, and Truth," Macmillan, New York, 1939.

How to Read Technical Statistical Analyses
Huck, S. W., W. H. Cormier, and W. G. Bounds, Jr.: "Reading Statistics and Research," Harper & Row, New York, 1974.

Review of Useful Mathematical Background
Auslander, L., et al.: "Mathematics Through Statistics," Williams & Wilkins, Baltimore, 1973.
Braverman, H.: "Reviewing Statistics," Robin Hill, New York, 1971.

SUPPLEMENTARY EXERCISES

1 The following data give the price per earnings (P/E) ratios of the top 60 companies in stock-market performance in 1977:

3	7	14	25	22	17
28	15	20	4	6	4
14	32	5	14	14	7
16	4	3	4	9	8
14	14	12	14	7	8
8	5	19	35	32	12
21	11	13	8	6	16
10	5	12	12	4	5
33	7	11	5	7	14
7.	7	17	27	19	5

(a) Construct a frequency distribution of the P/E ratios.
(b) Draw a frequency polygon which illustrates the distribution of part a.
2 As part of an experiment to study the growth rate of a new variety of plant, a botanist planted 100 germinated seeds and measured the heights of the plants 60 days later. The 100 heights, in inches, were as follows:

0.0	1.0	4.6	5.9	3.0	3.4	5.6	4.3	3.7	2.8
2.9	4.5	1.1	4.7	5.8	3.1	3.7	3.8	2.7	3.6
4.9	2.1	4.4	1.2	3.3	3.8	3.2	2.6	3.5	4.0
4.7	0.6	2.2	4.3	1.3	3.9	2.5	3.3	4.1	5.4
3.0	4.9	4.8	2.3	3.9	1.4	3.4	3.5	3.4	5.3
4.8	3.1	4.7	3.2	2.4	2.4	1.5	4.8	3.6	3.5
4.6	4.6	3.2	2.3	3.3	0.4	4.2	1.6	3.6	3.6
1.9	3.1	2.2	3.2	4.4	5.7	4.2	3.7	1.7	4.9
4.5	2.1	3.1	4.3	3.3	2.6	3.8	3.5	4.1	1.8
2.0	3.0	4.4	4.5	2.5	3.9	3.4	5.5	5.2	4.0

(a) Set up a frequency distribution of the heights.

(b) Construct a frequency polygon from the distribution of part a.

(c) Draw a histogram to illustrate the frequency distribution of part a.

3 The following data points from the files of a major life insurance company record the number of consecutive weeks that medical disability compensation was paid to 14 heart attack victims:

9	20
21	27
14	13
23	10
8	32
16	18
7	20

(a) Calculate the mean number of weeks that compensation was paid.

(b) Find the median number of weeks compensation was paid.

(c) Find the mode.

(d) Calculate the standard deviation of the number of weeks of paid compensation.

(e) Calculate the mean absolute deviation of the number of weeks of paid compensation.

4 A local bank having 12 branch offices is considering adopting a completely computerized payroll system. As part of an analysis of the operating costs of the system, the costs of running one aspect of the system are carefully observed for the 12 branches, with the following resulting data on costs per hour of operating the system:

24	20
14	24
14	14
14	20
12	14
7	12

(a) Calculate the mean operating cost for the 12-branch bank.

(b) Find the median operating cost.

(c) Find the modal operating cost.
(d) Calculate the standard deviation of the 12 operating costs.
(e) Calculate the mean absolute deviation of the costs.

5 In a biological study of the uniformity of growth of cherry tomato plants, the following data give the numbers of tomatoes harvested from 20 test plants after 90 days of growth:

3	8	11	17
22	24	19	14
9	1	17	18
12	9	12	25
22	17	13 ·	14

(a) Calculate the mean harvest.
(b) Draw a seesaw diagram of the data points, and indicate the location of the mean.
(c) Find the median harvest.
(d) Calculate the standard deviation.

6 The United States Smelting, Refining, and Mining Company had, until 5 years ago, a mill at Midvale, Utah, which produced a mean of 26,000 tons of lead concentrate per year over the past 36 years. The standard deviation of the annual production was 3500 tons. In at most how many of those 36 years could production have fallen below 15,500 tons?

7 An accountant commutes by automobile from his suburban home to his office in a downtown corporate headquarters. The daily trip takes on the average 24 minutes with a standard deviation of 4 minutes. If he were to leave his home regularly at 8:20 A.M.., at least how often would he manage to make it to the office in time for the staff summary meeting at 9 A.M.?

8 An automobile battery carrying a 24-month guarantee actually has a mean lifetime of 30 months with a standard deviation of 2 months.
(a) At most, how many of these batteries fail before 24 months have passed?
(b) The manufacturer is considering reducing the guarantee period in order to cut back on the expense of honoring the guarantee. At least how many of the batteries last longer than 20 months?

9 One local supermarket carries 20,000 different food items. With an eye toward both publicity and cost cutting, it announces that it will refuse to carry any item which has risen in price by more than 25% during the previous year. If food prices rose by an average of 12% one year with a standard deviation of 4%, at least how many of the original 20,000 items were still on the shelves the next year?

10 An urban planner's data indicate that on Los Angeles area freeways between 4 P.M. and 6 P.M. weekdays there are 400,000 vehicles. These vehicles, furthermore, spend a mean time of 45 minutes on the freeways with a standard deviation of 15 minutes. At most, how many vehicles spend 90 minutes or more on the freeways?

11 Over the past several years, the daily prime rate of interest at major New

York banks has averaged 6.2% with a standard deviation of .4%. At most, what proportion of the time has it exceeded 11.8%?

12 Several years ago, a group of urban planners constructed a model of a "planned" ideal city, of which one goal was that no more than 2% of the residents would have to commute longer than 20 minutes from home to work. When the actual city was completed recently, the inhabitants' mean travel time from home to work turned out to be 12 minutes with a standard deviation of 1 minute. Was the planners' goal achieved?

13 Insurance claims arrive at Buzzardbait National Life & Casualty's home office at a mean rate of 140 per week with a standard deviation of 12. At most, what proportion of weeks does the number of claims exceed 200?

14 Cars now on the road nationwide get 14 miles/gallon of gasoline on the average with a standard deviation of 3 miles/gallon. A plan now under study proposes to encourage engine efficiency by rebating all federal gasoline taxes to owners of those cars ranking among the top 10% in efficiency. If your car gets 25 miles to the gallon, would you qualify for a rebate?

15 The mean age of those eligible to vote in the United States is 40 years with a standard deviation of 10 years. At most, what percentage of eligible voters is in the age group 18 to 21?

16 In a college course on the construction and standardization of psychological tests, three exams are given. All three exams count equally toward the student's final grade, and nothing else is taken into consideration. The following table gives the mean and standard deviation of the scores of the entire class on each of the exams:

	First Exam	Second Exam	Third Exam
Class mean	65	74	71
Class standard deviation	4	6	12

One of the students, Xerxes, got a 57 on the first exam, a 76 on the second exam, and a 96 on the third, while another student, Yennie, obtained a 73 on the first, an 86 on the second, and a 70 on the third. Yennie was awarded B for the course, while Xerxes received only a C. Xerxes protests to the instructor, of course, because his average on the three exams was exactly the same as Yennie's, and so he feels that he deserves the same grade. Unfortunately for Xerxes, however, the instructor standardized each exam score with respect to the overall class performance on that exam before computing the student's grade.

(a) Calculate each of Xerxes' standardized scores, and compute the average of the three.

(b) Standardize each of Yennie's exam scores and average them.

(c) Explain why Yennie got a B and Xerxes a C.

17 Construct a set of eight data points having mean 12 and standard deviation 5.

(a) Standardize the data points.

(b) Verify by calculation that the z scores have mean 0 and standard deviation 1.

PROBABILITY
DISTRIBUTIONS

In the previous chapter, several techniques for describing a set of data were introduced in an attempt to organize the data into an understandable format. All the data sets which have been discussed in the text and the exercises, however, generally have had very different pictorial representations and numerical descriptions. It seemed that there was no way of telling in advance how the frequency polygon would look or what the mean and standard deviation would be. Yet, as it turns out, there are often discernible patterns common to sets of data derived from many different sources. In this chapter, our objective will be to further our understanding of statistical data by studying these general patterns of behavior of sets of data. If we can fit a particular data set with which we are working into one of these general patterns, then we will know more about its overall characteristics. We will then be able to say that it has all the properties of the general pattern to which it belongs, in addition to those special properties that can be discovered using the methods of the previous chapter. These general patterns of data behavior are called "probability distributions."

SECTION 2.A
PROBABILITIES AND PROPORTIONS

By the probability of an event, we usually mean the proportion of time that the event can be expected to occur in the long run. For a simple example, consider the tossing of a coin and the observation of which of the two sides, heads or tails, is facing upward when the coin lands. If the coin is tossed 10,000 times and falls heads on 3000 of these tosses, we are probably justified in asserting that the probability of heads for this particular coin is 3/10. By a "fair" coin, we mean a coin whose probability of heads is 1/2.

To give an example of how probabilities are determined from a set of data, let's take a look at the subject matter of Table 1.2, which we reproduce in briefer format as Table 2.1. We can see from Table 2.1 that on 23 of the 100 days involved in the study, the demand for electricity was between 26 and 28 million kilowatthours. We can rephrase this fact in probability language as follows: The study indicates that the probability of a daily demand for electricity between 26 and 28 million kilowatthours is 23/100, or .23. Using symbols, we can write

$$P(26 \leq D \leq 28) = .23$$

where $P(\ )$ means "probability of" and D stands for the daily demand for electricity in millions of kilowatthours. Technically speaking, D is called a "random variable." Random variables are variable quantities which take several possible values determined according to a probability distribution. The notation $26 \leq D \leq 28$ signifies that D is at least as large as 26 (since $26 \leq D$), but no larger than 28 (since $D \leq 28$); i.e., D is between 26 and 28, inclusive. Similarly, we can write the following remaining probabilities from Table 2.1:

$$P(20 \leq D \leq 22) = .04$$

$$P(23 \leq D \leq 25) = .08$$

TABLE 2.1
Frequency distribution of daily demand for electricity

Demand, millions of kilowatthours	Days
20–22	4
23–25	8
26–28	23
29–31	34
32–34	26
35–37	4
38–40	1
Sum = 100	

TABLE 2.2
Probability distribution of daily demand for electricity, *D*

a, millions of kilowatthours	*b,* millions of kilowatthours	$P(a \leq D \leq b)$
20	22	.04
23	25	.08
26	28	.23
29	31	.34
32	34	.26
35	37	.04
38	40	.01
		Sum = 1.00

$P(29 \leq D \leq 31) = .34$

$P(32 \leq D \leq 34) = .26$

$P(35 \leq D \leq 37) = .04$

$P(38 \leq D \leq 40) = .01$

The seven probabilities calculated above constitute what is called the "probability distribution of the daily demand for electricity in millions of kilowatthours" or, more simply, the "probability distribution of *D*." In Table 2.2, we formally display the probability distribution of *D*, which is merely a listing of the possible values of *D* together with the probabilities that *D* takes on those values.

In Table 2.3, we present the probability distribution for the toss of the coin discussed earlier in this chapter. Recall that the coin in question fell heads 3000 times in 10,000 tosses. (The coin fell tails, of course, the remaining 7000 times.)

In the remaining sections of this chapter, we will study in detail what are perhaps the two most useful probability distributions encountered in elementary statistical investigations: the binomial distribution and the normal distribution. Probabilities and proportions are discussed further, from another point of view, in Chap. 6.

TABLE 2.3
Probability distribution of outcomes of coin toss
(For a coin which fell "heads" 3000 times in 10,000 tosses)

Outcome	Probability
Heads	3/10
Tails	7/10
	Sum = 1

EXERCISES 2.A

1 From the data of Exercise 1.A.1, construct a probability distribution of the diameters of stones found in the delta region under study.
2 Using the data of Exercise 1.A.5, set up a probability distribution of the learning times.
3 Construct a probability distribution of the housing ratings which are given by the frequency distribution of Exercise 1.A.4.

SECTION 2.B
THE BINOMIAL DISTRIBUTION

The repeated tossing of a coin provides perhaps the simplest example of a process which generates binomially distributed data. There are only two possible outcomes, heads and tails, with $P(\text{heads}) = 1/2$ and $P(\text{tails}) = 1/2$ if the coin is fair. Furthermore, each toss of the coin is independent of all previous tosses.[1]

These are the characteristic properties of what is called a binomial experiment: (1) There are only two possible outcomes ("binomial," from Greek and Latin roots, means "having two names"), (2) the probabilities of the outcomes remain constant throughout the experiment (although they do not necessarily have to equal 1/2), and (3) each trial of the experiment is independent of all previous trials. Many examples of applied interest can be organized as binomial experiments, and therefore any resulting data taken in such a situation will have the binomial distribution. Consider the following few cases:

1 *A marketing survey* Several shoppers near the dairy case of a supermarket are asked the question, "Would you try a carton of chocolate flavored buttermilk if such a product were developed?" The two possible outcomes are "yes" and "no." Presumably each shopper's opinion would not depend on that of any other shopper, if the questioning were properly conducted. The probability of primary concern involved here is $p = P(\text{yes})$, the proportion of shoppers who would be willing to try the new product.
2 *A political poll* Registered voters are asked whether or not they favor repeal

---

[1]This statement about coin tossing has generated substantial philosophical controversy over the years, even though it can be proved by a simple experiment. The allegation has been made that, if a fair coin falls heads 10 times in a row, it is very likely to fall tails on the 11th toss. Much money has been lost by gambling on this untrue assertion. Even though the "law of averages" implies that a coin will fall heads about half the time and tails the other half, this does not place a moral obligation on the individual coin to pay back those 10 tails it owes. In fact, during the 11th toss, the coin itself does not remember that it has fallen heads on the last 10 consecutive tosses. All it knows while it is spinning in the air is that, because of its structure, it is equally likely to fall heads or tails on any given toss, including the 11th. The reader is invited to check this out by conducting a coin-tossing experiment. Take a handful of, say, 500 pennies and toss them in the air in a closed room having no holes in the floor. About half the coins will fall tails. Remove them, and toss the remaining ones a second time. Again, half will have fallen tails. Remove them and proceed. Each time only half the coins will fall tails, even though every coin tossed has fallen heads on all previous tosses.

of a certain tax law. The two possible outcomes are "for repeal" and "against repeal." If the poll is properly conducted, each voter's response should be independent of that of all other voters. The relevant probability is $p = P$(for repeal), the proportion of voters favoring repeal.

3 *A medical test* Certain patients are given blood tests for anemia. The two possible outcomes are "is anemic" and "is not anemic." We can be reasonably sure that whether or not a particular patient is anemic does not depend on the condition of any other patient. The probability here is $p = P$(is anemic), the proportion of patients showing anemia.

4 *A psychological cure* Some patients are given a new drug to cure mental depression. The two possible outcomes are "cured" and "not cured." Each patient's reaction to the drug ordinarily does not depend on the reaction of other patients. The probability $p = P$(cured) is the proportion of patients

5 successfully treated by the new drug.
   An agricultural experiment Seeds are planted in soil to which a proposed new chemical fertilizer, to aid in seed germination, has been added. The two possible outcomes are "germinated" and "did not germinate." The germination of each individual seed is not affected by the germination of any other seed. The probability of interest is $p = P$(germinated), the proportion of seeds which germinate with the aid of the new fertilizer.

6 *An actuarial problem* Several men aged 45 apply for 10-year term life insurance policies. The two possible outcomes are "died before age 55" and "lived to age 55." Whether or not one particular man dies before he reaches the age of 55 does not seem to be dependent on what happens to any other policyholder. The relevant probability is $p = P$(death before age 55), the proportion of 45-year-old men who die before they reach age 55.

All binomial experiments can be completely characterized by two numbers. The first is p, the probability of one of the outcomes; the other outcome must then have probability $1 - p$, for if P(heads) $= 1/2$, then P(tails) $= 1/2$, and if P(yes) $= 4/5$, then P(no) $= 1/5$. (The reason for this is that the probabilities must add up to 1.00 in order to represent 100% of the occurrences.) The second number is n, the number of times the experiment is repeated. All properties and mathematical formulas involved in the study of binomial experiments can be expressed in terms of the numbers n and p. Numbers which are characteristic of a situation are called the "parameters" of the problem. Much of statistics is concerned with the study of parameters, because if we can figure out what the parameters of a situation are, then we can determine everything else from them.[1]

Example 2.1 Coin Tossing Suppose we take three fair coins, toss them, and count the number of heads we have obtained. Here the parameters are $n = 3$ and $p = 1/2$, where $p = P$(heads). We can denote by H the total number of heads

[1]However, some situations do not easily lend themselves to the analysis of parameters. To answer these questions, we need an entirely different class of procedures called "nonparametric" statistics, a brief introduction to which appears in Chap. 8.

obtained with the three coins. The quantity H is a random variable having the possible values 0, 1, 2, and 3, because with three coins, we can get none, one, two, or three heads. (There is no way to get four or more heads with three coins.) Compute the following probabilities:

$P(H = 0)$ = probability of getting no heads with three coins

$P(H = 1)$ = probability of getting one head with three coins

$P(H = 2)$ = probability of getting two heads with three coins

$P(H = 3)$ = probability of getting three heads with three coins

SOLUTION   As described in Sec. 2.A, the probabilities above constitute the probability distribution of the number of heads obtained in the toss of three coins. In this case, we say that the number of heads has the binomial distribution with parameters $n = 3$ and $p = 1/2$.

Now, how do we compute these probabilities? In Table 2.4, we list all possible outcomes of the toss of the three coins. As can be seen from Table 2.4, there are eight possible ways the three coins, as a group, can turn up. Because $p = 1/2$, we know that for each of the three coins, heads and tails are equally likely. Therefore outcome 1 is just as likely to occur as outcome 8, and, in fact, all eight outcomes are equally likely to occur. This means that we can expect each outcome to occur about 1/8 or 12.5% of the time.

We are not directly interested in these outcomes, however, but in the number of heads obtained in each case. From Table 2.4, we can see that we obtain three heads from outcome 1, two heads from outcomes 2, 3, and 4, one head from outcomes 5, 6, and 7, and no heads from outcome 8. Therefore, we can expect to get three heads whenever outcome 1 occurs, namely, 12.5% of the time. We express this in mathematical symbols by writing $P(H = 3) = .125$. We can expect to get two heads the 12.5% of the time when outcome 2 occurs, the 12.5% of the time when outcome 3 occurs, and also the 12.5% of the time when outcome 4 occurs. Therefore, we would expect to obtain two heads a total of 12.5% + 12.5% + 12.5% = 37.5% of the time. In mathematical symbols, we write $P(H = 2) = .375$. By similar reasoning, we see that $P(H = 1)$

TABLE 2.4
Outcomes of the toss of three coins

Outcome	First Coin	Second Coin	Third Coin
#1	Heads	Heads	Heads
#2	Heads	Heads	Tails
#3	Heads	Tails	Heads
#4	Tails	Heads	Heads
#5	Tails	Heads	Tails
#6	Tails	Tails	Heads
#7	Heads	Tails	Tails
#8	Tails	Tails	Tails

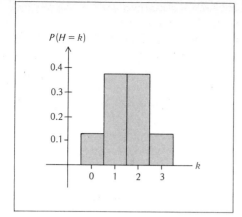

FIGURE 2.1

Histogram of the binomial distribution with
parameters $n = 3, p = \frac{1}{2}$.

$= .375$ and $P(H = 0) = .125$. By analogy with Table 2.2, we collect all these results
and display them in the probability distribution of Table 2.5. In Fig. 2.1, we
present the histogram of the probability distribution of Table 2.5.

In our calculation of the probability distribution of H, we relied very heavily on the fact that all eight outcomes of the coin tossing were equally likely
to occur. This situation was reflected in the value of the parameter p, which
was equal to 1/2. In the case $p \neq 1/2$, we no longer will have equally likely
outcomes. When $p > 1/2$, we are more likely to get a head than a tail, and
when $p < 1/2$, we are more likely to get a tail than a head. In such cases, the
calculation of the probability distribution is somewhat more difficult. Although there is a formula for calculating the various probabilities involved,
we have printed the results of those calculations in Table A.2 of the Appendix. We will therefore be able to study binomially distributed data by using
that table, for it lists probabilities for binomial distributions having various
numerical values for n and p.

TABLE 2.5
**Binomial distribution
with parameters
$n = 3, p = 1/2$**
(Number of heads, H)

k	$P(H = k)$
0	.125
1	.375
2	.375
3	.125
Sum = 1.000	

HOW TO USE THE TABLE OF THE BINOMIAL DISTRIBUTION

As an example of how to use the table, let's recompute the material in Table 2.5, using Table A.2 instead of the facts about coin tossing. The first fact we must note about Table A.2 is that the numbers there are values of $P(H \leq k)$ instead of $P(H = k)$. With this in mind, we reproduce in Table 2.6 the portion of Table A.2 dealing with $n = 3$ and $p = 1/2$ ($p = .50$ in the language of Table A.2). We then explain how to derive Table 2.5 from the information of Table 2.6. As has been just mentioned, the numbers in the right-hand column of Table 2.6 are probabilities of the type $P(H \leq k)$. This means that

$$.1250 = P(H \leq 0) = P(H = 0)$$

$$.5000 = P(H \leq 1) = P(H = 0) + P(H = 1)$$

$$.8750 = P(H \leq 2) = P(H = 0) + P(H = 1) + P(H = 2)$$

$$1.0000 = P(H \leq 3) = P(H = 0) + P(H = 1) + P(H = 2) + P(H = 3)$$

From the above information, we would like to compute the individual probabilities $P(H = 0)$, $P(H = 1)$, $P(H = 2)$, and $P(H = 3)$. First of all, since H can never be negative, $P(H \leq 0)$ is the same as $P(H = 0)$, so that

$$P(H = 0) = .1250 = .125$$

Then, because $P(H = 0) + P(H = 1) = .5000$, it follows that

$$P(H = 1) = .5000 - P(H = 0) = P(H \leq 1) - P(H \leq 0)$$
$$= .5000 - .1250 = .3750 = .375$$

Furthermore, because $P(H = 0) + P(H = 1) + P(H = 2) = .8750$, we have that

$$P(H = 2) = .8750 - [P(H = 0) + P(H = 1)]$$
$$= P(H \leq 2) - P(H \leq 1)$$
$$= .8750 - .5000 = .3750 = .375$$

TABLE 2.6
Table A.2 for $n = 3$, $p = .50$

n	k	$p = .50$
3	0	.1250
	1	.5000
	2	.8750
	3	1.0000

And, as $P(H = 0) + P(H = 1) + P(H = 2) + P(H = 3) = 1.0000$, we get

$$P(H = 3) = 1.0000 - [P(H = 0) + P(H = 1) + P(H = 2)]$$
$$= P(H \leq 3) - P(H \leq 2)$$
$$= 1.0000 - .8750 = .1250 = .125$$

We can summarize the above computations as follows:

$$P(H = 0) = P(H \leq 0) = .125$$
$$P(H = 1) = P(H \leq 1) - P(H \leq 0) = .375$$
$$P(H = 2) = P(H \leq 2) - P(H \leq 1) = .375$$
$$P(H = 3) = P(H \leq 3) - P(H \leq 2) = .125$$

It should now be observed that these are exactly the numbers appearing in the probability distribution of Table 2.5.

THE FORMULA FOR THE BINOMIAL DISTRIBUTION

Now that we have discussed how to use the table of the binomial distribution, let's briefly take a look at the formula for the binomial probabilities. If H is a binomial random variable having parameters n and p, we can calculate $P(H = k)$ as follows:

$$P(H = k) = \frac{n!}{k!(n - k)!} p^k (1 - p)^{n-k}$$

In this formula, the symbol $n!$ (pronounced "n-factorial") means $1 \times 2 \times \cdots \times n$, the product of all whole numbers between 1 and n, inclusive.[1] In particular, $3! = 1 \times 2 \times 3 = 6$ and $5! = 1 \times 2 \times 3 \times 4 \times 5 = 120$. The symbol p^k means $p \times p \times \cdots \times p$ k times, the product of k p's. For example, $(\frac{1}{2})^4 = \frac{1}{2} \times \frac{1}{2} \times \frac{1}{2} \times \frac{1}{2} = \frac{1}{16}$, and $(\frac{2}{3})^5 = \frac{2}{3} \times \frac{2}{3} \times \frac{2}{3} \times \frac{2}{3} \times \frac{2}{3} = \frac{32}{243} = .132$. In Example 2.1, we had $n = 3$ and $p = \frac{1}{2}$. Suppose we want to calculate $P(H = 1)$ using the formula. Then we would set $k = 1$, $n = 3$, and $p = \frac{1}{2}$. We get

$$P(H = 1) = \frac{3!}{1!(3 - 1)!} \left(\frac{1}{2}\right)^1 \left(1 - \frac{1}{2}\right)^{3-1} = \frac{3!}{(1!)(2!)} \left(\frac{1}{2}\right)^1 \left(\frac{1}{2}\right)^2$$

$$= \frac{6}{(1)(2)} \left(\frac{1}{2}\right)\left(\frac{1}{4}\right) = \frac{6}{16} = \frac{3}{8} = .375$$

Of course, this is exactly the same number we got using Table A.2.

[1] Due to certain technicalities, 0! turns out to be 1.

TABLE 2.7
Table A.2 for $n = 7$, $p = .60$

n	k	$p = .60$
7	0	.0016
	1	.0188
	2	.0963
	3	.2898
	4	.5801
	5	.8414
	6	.9720
	7	1.0000

As we continue to work with binomially distributed data in future portions of this text, we will refer to Table A.2 whenever numerical calculations are required.[1]

Example 2.2 A Political Poll Suppose it is really true that 60% of the voters are for repeal of a certain local tax law. If we were to select a random sample of seven voters, what would be the probability that a majority of the sample would be against repeal, thus making it appear that the antirepeal forces are in the lead?

SOLUTION Here we are dealing with a binomial experiment with $n = 7$. This can be viewed as analogous to tossing seven coins (representing the seven voters), each marked "for repeal" on one side and "against repeal" on the other. The fact that 60% of the voters are for repeal can be translated into coin-tossing language as the statement that $p = P(\text{for repeal}) = .60$. Therefore, if we abbreviate by F the number of voters that are *for* repeal, we see that F has the binomial distribution with parameters $n = 7$ and $p = .60$. In Table 2.7, we reproduce the relevant portion of the table of binomial distribution, Table A.2 of the Appendix. We have therefore completed the first part of the solution, which is to decide which portion of Table A.2 applies to the problem at hand.

For the second part of the solution, we have to decide how to make use of the information in Table A.2 in order to answer the question. To say that a majority of the seven voters is against repeal means that four or more are against repeal. This is equivalent to saying that three or fewer are for repeal, or that $F \leq 3$. It follows that

$$P(\text{a majority of the sample is against repeal}) = P(3 \text{ or fewer are for repeal})$$
$$= P(F \leq 3)$$

in view of the fact that F represents the number of voters that are for repeal. Now, as we have pointed out earlier, the probabilities in Table A.2 (and so Table 2.7) are exactly of the form $P(F \leq k)$. Therefore, in Table 2.7, to the right of 3 (of

[1]However, you are welcome to use the formula if you prefer.

refer to pg. 45. $p > .6$; more likely to be for a repeal.
$p < .4$; more likely to be against.
$p \leq .3$ are against.

TABLE 2.8
Binomial distribution with parameters $n = 7, p = .60$
(Number of voters for repeal, F)

k	$P(F \le k) - P(F \le k - 1) =$		$P(F = k)$
0	.0016 − .0000	=	.0016
1	.0188 − .0016	=	.0172
2	.0963 − .0188	=	.0775
3	.2898 − .0963	=	.1935
4	.5801 − .2898	=	.2903
5	.8414 − .5801	=	.2613
6	.9720 − .8414	=	.1306
7	1.0000 − .9720	=	.0280
		Sum =	1.0000

the k column), we find $P(F \le 3) = .2898$. Therefore, the probability is 28.98%, or almost 29%, that the majority of our sample will be against repeal even though 60% of the entire voting population are for repeal.

In Table 2.8 and Fig. 2.2, respectively, we record for illustrative purposes the probability distribution and the histogram of F. The calculations used to derive the information from Table 2.7 are also shown.

As the above example shows, there is more than one chance in four that a poll of seven voters would mislead the pollster into believing the one side was in the lead, even when the other side was really ahead 60% to 40%. How could this

FIGURE 2.2
Histogram of the binomial distribution with parameters $n = 7, p = .60$.

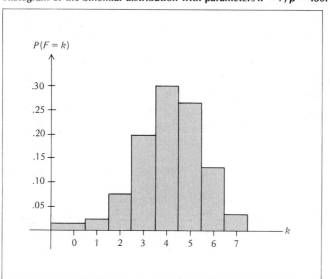

happen? The problem here is that seven voters do not constitute a sample large enough to inspire real confidence in the results. Suppose, for purposes of comparison, we had taken instead a sample of 19 voters instead of 7. Then, to assert that a majority is against repeal means that 9 or fewer are for repeal, namely $F \leq 9$. In Table A.2, for $n = 19$ and $p = .60$, we read that $P(F \leq 9) = .1861 = 18.61\%$. Therefore, working with 19 voters instead of 7, we would be able to reduce our chances of being seriously misled from 28.98% down to 18.61%. Our chances of error would naturally be even smaller were we to take a sample of 50 or 100 or more voters. While Table A.2 contains the binomial probabilities only for values of n up to $n = 20$, we will learn in the next two sections how to use Table A.3 for calculating binomial probabilities having the parameter n larger than 20.

Example 2.3 Effectiveness of a New Drug Suppose a new drug aimed at alleviating mental depression appears on the market. Prior to its licensing for public use, an intensive testing program has established that the drug is 80% effective. This means that it can be expected to alleviate mental depression in about 80% of the patients for whom it is prescribed. If we put 15 patients on the new drug, what are the chances that 12 or more of them will benefit from it?

SOLUTION   What we have here is a set of binomial data with parameters $n = 15$ and $p = .80$. If we denote by B the number of patients who benefit from the drug, we are interested in $P(B \geq 12)$, which is the same as $P(B > 11)$. Table A.2, as it is constructed, works better with $P(B > 11)$ than it does with $P(B \geq 12)$. In fact, the table contains neither of these items, but it does show that $P(B \leq 11) = .3518$ for $n = 15$, $p = .80$, and $k = 11$. This means that the chances are 35.18% that 11 or fewer patients benefit. It follows automatically that the chances are $100\% - 35.18\% = 64.82\%$ that more than 11 patients benefit. We can compactify this calculation by writing

$$P(B > 11) = 1 - P(B \leq 11) = 1 - .3518$$

$$= .6482 = 64.82\%$$

Therefore, if the drug is 80% effective, the probability is 64.82% that 12 or more of our 15 patients will benefit from it.

MEAN AND STANDARD DEVIATION OF THE BINOMIAL DISTRIBUTION

Before going on to the next section, we should call attention to a few more facts about the binomial distribution that we will find useful in our future work in statistics. These facts involve the mean and standard deviation of a binomial distribution which, as the material of Chap. 1 has shown, are important in understanding the behavior of random variables.

To introduce the concept of the mean of a probability distribution, let's take a look at some coin-tossing examples. Suppose we take a fair coin, having $p = P(\text{heads}) = .50$, and toss it 10 times. We would expect to get 5 heads in the 10 tosses. In fact, if several people got together and each tossed a fair coin 10 times, the number of heads tossed by each would average out to something near 5. It would therefore be reasonable to agree that the mean number of heads obtained in 10 tosses of a fair coin is 5.

Now, what would the situation be if we were to toss another coin which was not fair but instead had probability of heads $p = .80$? Then we would expect to obtain 8 heads in our 10 tosses, and so it would be reasonable to say that the mean number of heads in 10 tosses of the other coin is 8. For larger values of p, therefore, the mean number of heads would also be larger.

Suppose, on the other hand, we were to toss each coin 50 times instead of 10? Well, with a fair coin we would expect to get 25 heads, while with the coin having $p = .80$, we would expect 80% heads, or 40 heads out of the 50 tosses.

In general, suppose we take a coin having probability p of heads, and suppose we toss it n times. How many heads can we expect to get? We would expect that a proportion p of the n tosses would turn up heads. Therefore the number of heads expected would be $p \times n = np$ heads. Notice that this is exactly the procedure we followed in the specific cases above. In 10 tosses ($n = 10$) of a fair ($p = .50$) coin, we would expect $np = 10 \times .50 = 5$ heads, while in 50 tosses ($n = 50$) of a coin having $p = .80$, we would expect $np = 50 \times .80 = 40$ heads.

As Examples 2.1, 2.2, and 2.3 made clear, coin tossing is merely one special case of a large class of situations in which the data follow binomial distributions. Furthermore, whenever a random variable has a binomial distribution, all relevant probabilities can be determined from Table A.2 of the Appendix. This table is classified according to the parameters n and p of the random variable under study. All these random variables having the same n and p will also have the same mean, namely, np. This is true regardless of whether they are the mean number of heads tossed, the mean number of voters for repeal, the mean number of patients benefitting, or the mean of any other binomially distributed random variable having parameters n and p.

We are therefore justified in using the formula

$$\star \quad \mu = np \tag{2.1}$$

to calculate the mean of any set of binomially distributed data having parameters n and p. An analogous argument would show that we should use the formula

$$\star \quad \sigma = \sqrt{np(1 - p)} \tag{2.2}$$

to calculate the standard deviation of a set of binomial data having parameters n and p. We will have occasion to use these formulas for μ and σ in applied

TABLE 2.9
Calculation of μ and σ for some binomial distributions

n	p	$\mu = np$	$1 - p$	$np(1 - p)$	$\sigma = \sqrt{np(1 - p)}$
10	.30	3	.70	2.1	1.45
10	.50	5	.50	2.5	1.58
10	.80	8	.20	1.6	1.26
50	.30	15	.70	10.5	3.24
50	.50	25	.50	12.5	3.54
50	.80	40	.20	8.0	2.83

contexts from time to time. In Table 2.9, we present some sample calculations of μ and σ.

EXERCISES 2.B

1 Suppose we take four coins, toss them, and count the number of heads obtained, denoting the resulting number of heads by the random variable H.
   (a) What are the parameters of the binomial random variable H?
   (b) What are the possible values of H?
   (c) Construct the probability distribution of H.

2 An executive in charge of new product development at a local dairy feels that about 10% of the population would be willing to try a carton of chocolate flavored buttermilk, while the other 90% would not be at all interested. To support her feeling with some observable facts, she conducts a survey at a local supermarket and asks each of 15 persons whether or not they would be willing to try a carton. She denotes by Y the random variable indicating the number of persons responding "yes" to the question.
   (a) What are the parameters of the random variable Y?
   (b) What are the possible values of Y?
   (c) Calculate the mean of Y.
   (d) Calculate the standard deviation of Y.
   (e) Construct the probability distribution of Y.

3 Twelve patients manifesting certain special symptoms are given blood tests for anemia. Usually, 20% of patients having those symptoms are judged by the test to have some form of anemia. We denote by A the random variable indicating the number of persons of the 12 tested who are judged to have some form of anemia.
   (a) State the parameters of the random variable A.
   (b) List the possible values of A. (12) $A = 0$ thru 12 n
   (c) Calculate the mean of the random variable A.
   (d) Calculate the standard deviation of the random variable A.
   (e) Set up the probability distribution of A. figure out each probabilities + draw

4 The manufacturer of a new chemical fertilizer guarantees that with the aid of his fertilizer, 80% of the seeds planted will germinate. To check on his assertion, we plant 17 seeds as directed and denote by G the number which germinate.

(a) What should the parameters of G be?
(b) What are the possible values of G?
(c) What is the probability that fewer than 10 of our 17 seeds will germinate?
(d) Construct the probability distribution of G.

5 Twenty men aged 45 apply for a 10-year term life insurance policy. In order to know how much to charge for the policy, the issuing company needs to be able to estimate how many of the applicants will die before age 55, in which case the company will have to pay off on the policy. Actuarial studies indicate that of all 45-year-old men in the applicants' occupational grouping, 5% will die before they reach age 55.
(a) Find the parameters of the random variable D which denotes the number of applicants who die before they reach age 55.
(b) What is the probability that none of the 20 applicants will die before age 55, making it a very profitable deal for the insurance company?
(c) Find the probability that exactly 1 of the 20 will die during the 10-year term.
(d) What are the chances that more than 5 of the 20 applicants will die before age 55?
(e) What is the probability that fewer than 10 of the applicants will die during the specified period?
(f) Calculate the probability that more than 5 but fewer than 10 applicants will not make it to age 55.

6 After reading a news reporter's allegation that 65% of the membership of the U.S. House of Representatives favors a certain controversial change in the banking laws, a staff assistant to the major lobbying group opposing the change conducts a quick "straw poll" of 14 randomly selected House members.
(a) If the reporter's story is correct, what are the chances that fewer than half of the 14 members surveyed would favor the change?
(b) What is the probability that exactly half of those surveyed favor the change?
(c) What would be the probability that more than 12 of the 14 members polled would favor the change?
(d) What are the chances that more than half but fewer than three-fourths of those questioned would favor the change?

SECTION 2.C
THE NORMAL DISTRIBUTION

There are two fundamental sources from which normally distributed data arise: repetition of the same physical process over and over again (such as the machine filling of 8-ounce cans of tomato sauce) and artificial standardization of test scores (such as scores on the Graduate Record Examination, the Admission Test for Graduate Study in Business, etc.). A set of data can be often recognized as having an approximately normal distribution if its frequency polygon is a bell-shaped curve, as illustrated in Fig. 2.3. The word "normal" comes from

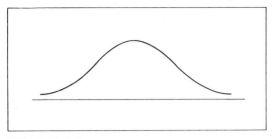

FIGURE 2.3
A "bell-shaped" curve.

historical usage and is not meant to imply that other distributions are "abnor-mal" in any sense. As examples of the occurrence of normally distributed sets of data in situations of applied interest, consider the following:

1 *Contents of cereal boxes* It is a well-known fact that Corn Puffies are packed by machine into 9-ounce boxes. The great majority of boxes contain amounts very close to 9 ounces, a little more or a little less, but every once in a while a box shows up having contents of quite a bit more or quite a bit less. Contents in ounces of boxes of Corn Puffies are normally distributed with mean $\mu = 9$ and standard deviation $\sigma = .01$. The fact that the standard deviation is so small, relatively speaking, means that virtually all the boxes have almost exactly 9 ounces of cereal, while very, very, very few differ appreciably from it. The intense concentration of data points near 9 results in the tall and thin bell-shaped frequency polygon of Fig. 2.4.

2 *Heights and weights of persons* People living in a certain area, whether the specified area is a political subdivision such as a city, state, or nation or a geographical subdivision such as a valley, plain, or delta region, have nor-

FIGURE 2.4
Frequency polygon of normally distributed con-tents of Corn Puffies boxes.

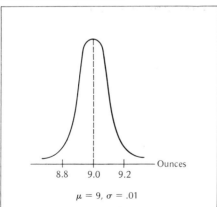

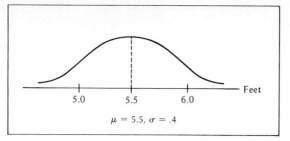

FIGURE 2.5
Normal curve of people's heights.

mally distributed heights and weights. Other descriptive measures such as shirt or shoe sizes are also normally distributed. A particular group of people may have mean height $\mu = 5.5$ feet with a standard deviation $\sigma = .4$ feet. The degree of concentration near the mean would therefore not be as extreme as in the case of the Corn Puffies data of the previous example, which had a smaller σ. It follows logically that the lessened degree of concentration near the mean would result in a flatter bell-shaped curve. The normal curve of heights appears in Fig. 2.5.

3 *Scores on standardized tests*   Scores of all students taking the verbal portion of the Graduate Record Examination (GRE) are artificially standardized so as to have a normal distribution with mean $\mu = 500$ and standard deviation $\sigma = 100$. The purpose of such standardization is to allow persons who use the results of such tests to be better able to interpret the results in terms of the relative performances of several candidates. The frequency polygon of the GRE scores is graphed in Fig. 2.6.

Each set of normally distributed data can be completely characterized by two numbers, its mean μ and its standard deviation σ. As explained in Chap. 1, μ specifies the location of the mean while σ measures the extent to which the data points are concentrated near the mean. The numbers μ and σ, since they completely describe the data, are the parameters of the normal distribution.

FIGURE 2.6
Frequency polygon of GRE scores.

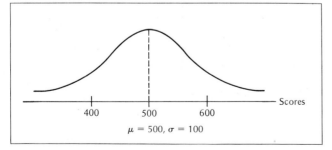

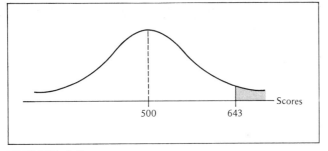

FIGURE 2.7
Proportion of GRE scores above 643.

Example 2.4 The Graduate Record Examination As we know, verbal scores on the GRE are normally distributed with mean $\mu = 500$ and standard deviation $\sigma = 100$. If a student scored 643, what percentage of those who took the test ranked higher than she did?

SOLUTION The graph of the normal distribution involved in this example is shown in Fig. 2.7. The problem is to calculate the proportion of data falling in the shaded region to the right of the number 643. There is a formula for calculating the proportions of a set of normally distributed data that fall in various intervals. Unfortunately, however, to be able to read this formula out loud, let alone to know how to use it, requires a considerable knowledge of calculus. Fortunately, we have in Table A.3 of the Appendix a table of the various probabilities connected with a set of normally distributed data. Before we go any further, therefore, it is necessary to discuss how to use Table A.3. It should also be pointed out here that there are some other types of probability distributions whose frequency polygons appear bell-shaped, but these cannot be the normal distribution unless their probabilities are consistent with those of Table A.3.

HOW TO USE THE TABLE OF THE NORMAL DISTRIBUTION

First of all, as the commentary preceding Table A.3 indicates, it is a table of the standard normal distribution. What does this mean? If you recall the discussion of standard scores, or z scores, in Sec. 1.F, you may have noted that we can calculate the z scores of any set of data and that these z scores have a probability distribution with mean $\mu = 0$ and standard deviation $\sigma = 1$. The operation of calculating the z scores of a set of data is called "standardizing" the data. If we calculate the z scores of a normally distributed set of data, namely, if we standardize a set of normally distributed data, we will obtain a set of normally distributed data having mean $\mu = 0$ and standard deviation $\sigma = 1$. The resulting normal distribution having the parameters $\mu = 0$ and $\sigma = 1$ is called the "standard" normal distribution, so that the z scores of any normal distribution have the standard normal distribution.

It would be grossly impractical to print tables of normal distributions for *every possible combination* of the numbers μ and σ, for this would require an extremely large number of pages, infinitely many, to be sure. Fortunately, however, in problems dealing with any normal distribution, it is sufficient to use the table of the standard normal distribution, merely by working with the z scores instead of the original data points (raw scores). To see how this is done, let's return to the GRE example.

What we want to know is the proportion of scores falling above 643. This proportion will be exactly the same as the proportion of z scores falling above the z score corresponding to 643. Now, as you may recall from Sec. 1.F, the z score corresponding to the number x is given by the formula

$$z = \frac{x - \mu}{\sigma}$$

Therefore the z score corresponding to 643 is

$$z = \frac{643 - 500}{100} = \frac{143}{100} = 1.43$$

because here $\mu = 500$ and $\sigma = 100$.

The original and standardized normal distributions involved are illustrated in Fig. 2.8. As the pictures indicate, we can determine the proportion of candidates scoring above 643 on the verbal portion of the GRE merely by calculating the proportion of a normal distribution which falls *above* the z score 1.43. Table A.3 is basically a list of z scores, together with the proportion of a set of standard normal data falling *below* the z score. Looking for the z score $z = 1.43$ in Table A.3, we find next to it the proportion .9236, which indicates that 92.36% of a standard normal distribution lies *below* the z score $z = 1.43$. We reproduce the relevant portion of Table A.3 in Table 2.10. It follows, then, that $100\% - 92.36\% = 7.64\%$ of the distribution must lie *above* $z = 1.43$. Returning to the GRE data,

FIGURE 2.8
Original and standardized GRE scores.

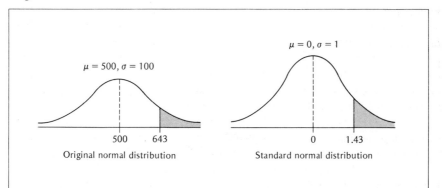

$\mu = 500, \sigma = 100$

500   643
Original normal distribution

$\mu = 0, \sigma = 1$

0   1.43
Standard normal distribution

TABLE 2.10
A small portion of Table A.3

z	Proportion
1.43	.9236

we can conclude that 7.64% of all candidates taking the GRE score above 643. A student scoring 643 can therefore be sure of easily qualifying for the top 10% of all students taking the GRE since only 7.64% could have scored higher.

Example 2.5 The Graduate Record Examination (Another View) Now that we know that only 7.64% of students score above 643, it would be useful and interesting to know what score will qualify a student for the top 5% of all candidates.

SOLUTION   As it turns out, we also can answer this question by using Table A.3. A graph illustrating the situation appears in Fig. 2.9, with the letter x denoting the (unknown) test score which separates the lower 95% of candidates from the upper 5%. Because $\mu = 500$ and $\sigma = 100$, the z score of the unknown x can be written as

$$z = \frac{x - 500}{100}$$

Because of this algebraic relationship between x and z, we will be able to compute x if we know the numerical value of z. As the picture of the standard normal distribution of Fig. 2.9 shows, 5%, or .0500, of the z scores will be above z, since 5% of the original scores are above x. This implies that 95%, or .9500, of the z scores will be below z. From Table A.3, we can find the particular z which has

FIGURE 2.9
Original and standardized GRE scores.

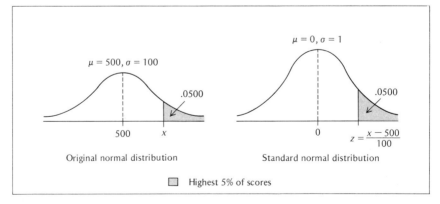

TABLE 2.11
A small portion of Table A.3

z	Proportion
1.64	.9495
1.65	.9505

95% of the z scores below it. All we have to do is find the *proportion* in Table A.3 which is closest to .9500, and then we read off the corresponding value of z.

The relevant portion of Table A.3 is reproduced in Table 2.11. Unfortunately, our table is not sufficiently detailed as to include the number .9500 exactly. As Table 2.11 shows, there are two proportions, .9495 and .9505, which are equally close to .9500. In a case like this, it is customary to choose the z score which is closer to 0; so let's say that the z score in question is $z = 1.64$. For purposes of comparison, the graph of the standard normal distribution illustrating the position of $z = 1.64$ appears in Fig. 2.10 together with the standard normal distribution of Fig. 2.9. From the pictures in Fig. 2.10, we note the following two facts:

1 5% of a standard normal distribution lies to the right of

$$z = \frac{x - 500}{100}$$

2 5% of a standard normal distribution lies to the right of

$$z = 1.64$$

A comparison of the two pictures in Fig. 2.10 then shows that

$$\frac{x - 500}{100} = 1.64$$

FIGURE 2.10
A comparison of standard normal distributions.

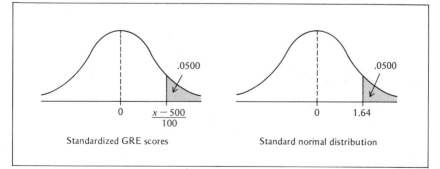

Standardized GRE scores	Standard normal distribution

Successive algebraic operations then yield that

$$x - 500 = (1.64)(100) = 164$$

and so

$$x = 164 + 500 = 664$$

We have therefore found the numerical value (664) of

x = the test score which seperates the lower 95% of candidates from the upper 5%

It follows that a score above 664 on the verbal portion of the GRE will put a student in the top 5% of all persons taking the test.

Example 2.6 Suit Sizes   Men living in a certain urban area have mean suit size 38 with a standard deviation of 2.5. Furthermore, their suit sizes are normally distributed. If a small ready-to-wear clothing store plans to begin business with an initial stock of 500 suits, how many of these should be of sizes between 35 and 40?

SOLUTION   First of all, we should remark that men's *actual* suit sizes are *continuous*, while ready-to-wear suits are manufactured only in *discrete* sizes. For example, an individual man may have size 35.8430067, but he will have to buy a size 36 suit even though it will not fit him exactly. The normal distribution is a *continuous* distribution because Table A.3 admits all numbers as possible values of z. For example, to compute the proportion of men who are fitted with size 36 suits, we would find the proportion of men having actual physical sizes between 35.5 and 36.5. We are in fact computing the probability that a normal random variable with mean 38 and standard deviation 2.5 will have a value between 35.5 and 36.5. The binomial distribution, on the other hand, is a *discrete*[1] distribution because only certain numbers are admitted as possible values. For example, if we toss a coin three times, we can get only 0, 1, 2, or 3 heads. There is no way of getting 2.5 or 2.6768 as a possible value of a binomial random variable.

With the above facts in mind, we see that we need to compute the proportion of men having *actual* suit sizes between 34.5 and 40.5. If we do this, we will know what proportion of men will need suits labeled between 35 and 40. The diagrams appear in Fig. 2.11, where the z score of 34.5 is

$$z(34.5) = \frac{34.5 - \mu}{\sigma} = \frac{34.5 - 38}{2.5} = \frac{-3.5}{2.5} = -1.40$$

---

[1] The spelling of the word "discrete," as opposed to "discreet," does not necessarily imply that the binomial distribution is indiscreet. It merely implies that it is not indiscrete; namely, it is not continuous.

bar

60   ESSENTIALS OF STATISTICS

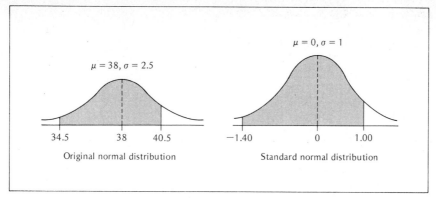

FIGURE 2.11
Original and standardized suit sizes.

and the z score of 40.5 is

$$z(40.5) = \frac{40.5 - \mu}{\sigma} = \frac{40.5 - 38}{2.5} = \frac{2.5}{2.5} = 1.00$$

From Table A.3, or from the portion of it appearing in Table 2.12, we see that 8.08% of the z scores lie below −1.40, while 84.13% lie below 1.00. Of the 84.13% which lie below 1.00, then, 8.08% also lie below −1.40, and so the difference 84.13% − 8.08% = 76.05% falls between −1.40 and 1.00. Returning to the original normal distribution of suit sizes, we conclude that 76.05% of men have sizes between 34.5 and 40.5 and would therefore be fitted with suits labeled 35 to 40. Of the initial stock of 500 suits, then, the store should order 76.05% or 380 suits of sizes between 35 and 40.

Example 2.7 Child Development Experimental results in child development show that ages at which children learn to walk are normally distributed with mean 11.2 months and standard deviation .4 months. One theory classifies as "early walkers" those children aged among the youngest 2% according to the time they learn to walk. If a child is to qualify as an early walker, by what age must he or she learn to walk?

SOLUTION   If we denote by x the (unknown) age which qualifies a child as an early walker, then the definition of early walker implies that 2% of the normally

TABLE 2.12
A small portion of Table A.3

z	Proportion
−1.40	.0808
1.00	.8413

TABLE 2.13
A small portion of Table A.3

z	Proportion
−2.06	.0197
−2.05	.0202

distributed data falls below (i.e., to the left of) *x*. This means that 2% of a standard normal distribution lies to the left of

$$z = \frac{x - \mu}{\sigma} = \frac{x - 11.2}{.4}$$

From Table A.3, or the remnant of it appearing in Table 2.13, we see that the *z* score of *x* must be −2.05 (because .0202 is closer to .0200 than .0197 is). Therefore

$$\frac{x - 11.2}{.4} = -2.05$$

and some algebraic transformations yield that, successively,

$$x - 11.2 = (-2.05)(.4) = -.82$$

and

$$x = -.82 + 11.2 = 10.38$$

FIGURE 2.12
Original and standardized ages of learning to walk (months).

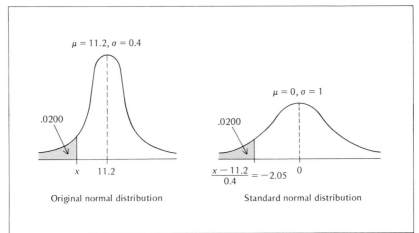

Original normal distribution	Standard normal distribution

It follows that children who learn to walk at age 10.38 months or earlier are classified as early walkers. The diagrams of the relevant normal distributions appear in Fig. 2.12.

EXERCISES 2.C

1 The length of time it takes grade-school pupils to complete a standard nationwide arithmetic progress test is a normally distributed random variable having mean 58 minutes and standard deviation 9.5 minutes. The educational psychologist who designed the exam wants 90% of the pupils who take it to complete it within a specified time allowed for the test. How much time should be allowed?

2 Records of the Wetspot Washing Machine Company indicate that the length of time their washing machines operate without requiring repairs is normally distributed with mean 4.3 years and standard deviation 1.6 years. The company repairs free any machine which fails to work properly within 1 year after purchase. What percentage of Wetspot machines require these free repairs?

3 The mathematics portion of the nationally administered Graduate Record Examination is graded in such a way that the scores are normally distributed with mean 500 and standard deviation 100.
(a) What proportion of applicants score 682 or higher?
(b) What proportion score between 340 and 682?

4 Men's shoe sizes nationwide are normally distributed with mean 10.5 and standard deviation 1.2. What proportion of men have shoe sizes between 8.25 and 12.25?

5 Lamps used in residential area street lighting are constructed to have a mean lifetime of 400 days with a standard deviation of 30 days. Furthermore, their lifetimes are normally distributed.
(a) What percentage of such lamps last longer than 1 year (365 days)?
(b) What percentage last between 375 and 425 days?
(c) What percentage last longer than 480 days?

6 Rainfall for the month of March in a geographic area in South America is normally distributed with mean 3.6 inches and standard deviation 1.0 inch. What proportion of years have a total rainfall in March exceeding 2.0 inches but not exceeding 3.0 inches?

7 One psychology instructor makes sure that every set of examination scores in her course is normally distributed. Then she assigns A grades to the top 15% of the scores, while the bottom 15% are awarded F's. A particular set of exam scores turns out to have a mean of 60 and a standard deviation of 16. What score is the dividing line between A and B, and what score between D and F?

8 The New York Stock Exchange is open 250 days each year. Closing prices of shares of Brownline Copper Co. (BCC) averaged $21\frac{1}{2}$ in the past year with a standard deviation of 4.3. Furthermore, BCC stock closed above $32\frac{1}{4}$ on 28 different days during the year. Were closing prices of BCC shares normally distributed?

9 A small auto-parts company markets a 12-volt battery which, according to statistical studies, lasts an average 1200 days with a standard deviation of 50 days. The lifetimes of the batteries were also shown to be approximately normally distributed. The company would like to put a guarantee on its product so that no more than 10% of the batteries will fail before the guarantee runs out. For how many days should the battery be guaranteed?

10 The growing season, planting to harvesting, of a certain variety of tomato is approximately normally distributed with a mean of 100 days and a standard deviation of 10 days. What percentage of such tomatoes can be harvested within 75 days of planting?

SECTION 2.D
THE NORMAL APPROXIMATION TO THE BINOMIAL DISTRIBUTION

Table A.2, the table of probabilities of the binomial distribution, contains information only for situations when the parameter n has a value of 20 or less. Because of this restriction, we have been unable up to now to calculate, for example, the probability of tossing 40 or fewer heads in 100 tosses of a fair coin, because $n = 100$ for this problem. Fortunately, a result from advanced mathematics called the central limit theorem (to be discussed in greater detail in Sec. 3.B) gives the surprising information that the normal distribution can be used to find that probability.

What the central limit theorem says in regard to this problem is that when $np(1 - p)$ is greater than 5 in magnitude, a binomial random variable with parameters n and p can be considered normally distributed with parameters $\mu = np$ and $\sigma = \sqrt{np(1 - p)}$. In this form, the central limit theorem is referred to as "the normal approximation to the binomial distribution." In the case of 100 tosses of a fair coin, the parameters of the binomial random variable H (the number of heads obtained in the 100 tosses) are $n = 100$ and $p = .50$. Therefore $np(1 - p) = 100(.50)(1 - .50) = 50(.50) = 25 > 5$, so that the normal approximation to the binomial distribution applies to this problem. Therefore H can be considered as a normal random variable with parameters $\mu = np = 100(.50) = 50$ and

$$\sigma = \sqrt{np(1 - p)} = \sqrt{100(.50)(1 - .50)} = \sqrt{25} = 5$$

Furthermore, since H is a discrete random variable, as discussed in Example 2.6 above, we actually should calculate the probability that $H < 40.5$, namely, that we toss fewer than 40.5 heads, if we are looking for the probability of tossing 40 or fewer heads in 100 tosses of the coin. The bell-shaped curves for the coin-tossing situation are presented in Fig. 2.13. The z score of 40.5 is

$$z = \frac{40.5 - \mu}{\sigma} = \frac{40.5 - 50}{5} = \frac{-9.5}{5} = -1.90$$

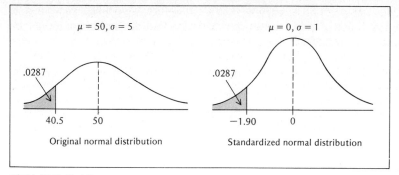

$\mu = 50, \sigma = 5$

$\mu = 0, \sigma = 1$

.0287

.0287

40.5   50

-1.90 0

Original normal distribution

Standardized normal distribution

FIGURE 2.13
Original and standardized number of heads tossed.

and Table A.3 indicates that the proportion of a standard normal distribution falling below -1.90 is .0287. Therefore the probability of tossing 40 or fewer heads in 100 tosses of a fair coin is only 2.87%, or 287 chances out of 10,000.

Example 2.8 A Political Poll (This example is a modified version of Example 2.2.) Suppose it is really true that 60% of the voters are for repeal of a certain local tax law. If we were to select a random sample of 70 voters, what is the probability that a majority of the sample (36 or more voters) would be against repeal, thus making it appear that the antirepeal forces are in the lead?

SOLUTION   Here we are dealing with a binomial random variable F, the number of voters that are *for* repeal, having parameters $n = 70$ and $p = .60$. The normal approximation is valid because $np(1 - p) = 70(.60)(1 - .60) = 42(.40) = 16.8$ is greater than 5. The probability that a majority of the sample of 70 is against repeal is the same as the probability that F is fewer than 34.5, and this latter probability is the same as the proportion of a standard normal distribution that falls below

$$z = \frac{34.5 - \mu}{\sigma}$$

where

$$\mu = np = 70(.60) = 42$$

$$\sigma = \sqrt{np(1 - p)} = \sqrt{16.8} = 4.1$$

We then have that

$$z = \frac{34.5 - 42}{4.1} = \frac{-7.5}{4.1} = -1.83$$

From Table A.3, we find out that the probability that a standard normal random variable falls below −1.83 is .0336, or 3.36%. Therefore, if we take a sample of 70 voters, our chances of being misled into believing that the antirepeal forces are in the lead are only 3.36%.

Referring back to Example 2.2 of Sec. 2.B, we note that our chances of being so misled were 28.98% when we used a sample of 7 voters, and 18.61% when we used a sample of 19 voters. As we have just shown, the probability of being misled drops considerably, to 3.36%, when a sample of 70 voters is used. This demonstrates convincingly that by increasing the sample size we can obtain a substantial decrease in our probability of error.

EXERCISES 2.D

1 As part of a coin-tossing experiment to check on the laws of probability (presumably to find out if they should be repealed), a fair coin is tossed several times, and the number of heads obtained is carefully counted and denoted by H.
   (a) If the coin is tossed 4 times, find the probability that H, the number of heads, will be at least 2 and no larger than 3.
   (b) If the coin is tossed 40 times, find the probability that H will be at least 20 and no larger than 30.
   (c) If the coin is tossed 400 times, find the probability that H will be at least 200 and no larger than 300.
   (d) If the coin is tossed 4000 times, find the probability that H will be at least 2000 and no larger than 3000.
2 The manufacturer of a new chemical fertilizer guarantees that with the aid of her fertilizer, 80% of the seeds planted will germinate. If we believe the manufacturer's assertion,
   (a) With what probability can we expect more than 13 of 17 seeds planted to germinate?
   (b) With what probability can we expect more than 130 of 170 seeds planted to germinate?
   (c) With what probability can we expect more than 1300 of 1700 seeds planted to germinate?
   (d) With what probability can we expect more than 13,000 of 17,000 seeds planted to germinate?
3 Actuarial studies indicate that of all 45-year-old men in a certain occupational grouping, 5% will die before they reach age 55. As part of the process of deciding whether or not to offer a 10-year term group life insurance policy to all 45-year-old men in that occupational grouping, an insurance company has to know the answers to the following questions:
   (a) If 20 men apply for the policy, what are the chances that more than 2 of them will die before reaching age 55?
   (b) If 200 men apply, what are the chances that more than 20 of them will die before age 55?
   (c) If 2000 men apply, what are the chances that more than 200 of them will die before age 55?

(*d*) If 20,000 men apply, what are the chances that more than 2000 of them will die before age 55?

4 A news reporter alleges that 65% of the membership of the 50 state legislatures favors a certain controversial change in the banking laws. If the reporter's story is correct,

(*a*) Find the probability that no more than half of 10 randomly selected legislators support the change.

(*b*) Find the probability that no more than half of 100 randomly selected legislators actually support the change.

(*c*) Find the probability that no more than half of 400 randomly selected legislators actually support the change.

5 A new drug aimed at alleviating mental depression seems to be 80% effective, according to an intensive prelicensing testing program.

(*a*) If the new drug is prescribed for 15 patients, with what probability can we expect fewer than 11 to benefit from it?

(*b*) If the drug is prescribed for 150 patients, how likely is it that fewer than 110 will benefit?

(*c*) Of 1500 patients taking the drug, what are the chances that fewer than 1100 will benefit from it?

(*d*) With 15,000 patients being given the new drug, what is the probability that fewer than 11,000 will benefit from it?

SECTION 2.E
PERCENTAGE POINTS OF THE NORMAL DISTRIBUTION

In many problems of applied interest that we will encounter in the rest of the book, we will be working with various quantities which have the standard normal distribution. To be better able to specify the location on the normal distribution of such quantities, we will introduce a system of labeling the distribution.

Using the Greek letter α (pronounced "alpha") to represent a proportion of data, we will use the symbol $z_{\alpha/2}$ to signify the z value of a standard normal distribution which has, to its right, a fraction $\alpha/2$ of the data. The meaning of $z_{\alpha/2}$ is illustrated in Fig. 2.14. Because $z_{\alpha/2}$ has a proportion $\alpha/2$ of the data to its right, it must have the remainder, a proportion $1 - \alpha/2$ to its left. For example, taking

FIGURE 2.14
The meaning of $z_{\alpha/2}$.

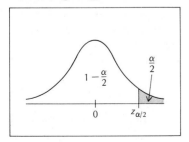

TABLE 2.14
A small portion of Table A.3

z	Proportion
1.96	.9750

TABLE 2.15
A small portion of Table A.3

z	Proportion
1.64	.9495
1.65	.9505

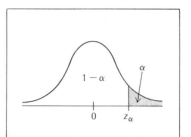

FIGURE 2.15
The meaning of z_α.

$\alpha = .05$, we have $\alpha/2 = .025$. Here $z_{\alpha/2}$ is $z_{.025}$, the z score which has .025, or 2.5%, of the data to its right and therefore .975, or 97.5%, to its left. We can find the numerical value of $z_{.025}$ by using Table A.3 of the Appendix because $z_{.025}$ is the z score corresponding to the proportion .9750. The portion of Table A.3, the part reproduced in Table 2.14, shows that $z_{.025} = 1.96$.

Suppose for $\alpha = .05$, we needed to know z_α, the z score which has a proportion α of the data to its right. As illustrated in Fig. 2.15, this means that a fraction $1 - \alpha$ of a set of standard normal data lies to the left of z_α. When $\alpha = .05$, z_α is $z_{.05}$, which has 5% of the data to its right and $1 - \alpha = 1 - .05 = .95 = 95\%$ to its left. From the portion of Table A.3 which is reproduced in Table 2.15, we see that the exact value of $z_{.05}$ lies somewhere between 1.64 and 1.65. Since we have agreed to choose the z score which is closer to 0 in cases like this, we shall set $z_{.05} = 1.64$.

In Table 2.16, we list z_α and $z_{\alpha/2}$ for several selected values of α. The numbers z_α correspond to the numbers $1 - \alpha$ of Table A.3, while the $z_{\alpha/2}$'s correspond to the $1 - \alpha/2$'s. In Fig. 2.16, we illustrate the position of various percentage points of the normal distribution. We will make extensive use of percentage points throughout the rest of the book, beginning with the next chapter.

TABLE 2.16
Some selected percentage points of the normal distribution

α	$1 - \alpha$	z_α	$\alpha/2$	$1 - \alpha/2$	$z_{\alpha/2}$
.20	.8000	.84	.10	.9000	1.28
.10	.9000	1.28	.05	.9500	1.64
.05	.9500	1.64	.025	.9750	1.96
.025	.9750	1.96	.0125	.9875	2.24
.02	.9800	2.05	.01	.9900	2.33
.01	.9900	2.33	.005	.9950	2.57
.005	.9950	2.57	.0025	.9975	2.81

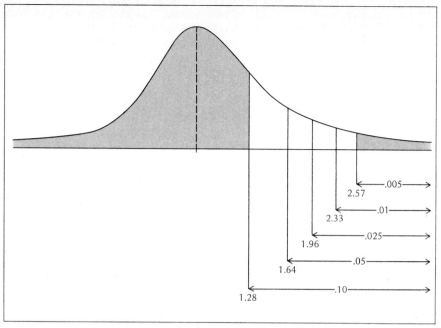

FIGURE 2.16
Some percentage points of the normal distribution.

EXERCISES 2.E

1 For each of the following numerical values of α, find z_α:
   (a) $\alpha = .30$ (e) $\alpha = .015$
   (b) $\alpha = .15$ (f) $\alpha = .008$
   (c) $\alpha = .075$ (g) $\alpha = .002$
   (d) $\alpha = .03$ (h) $\alpha = .001$
2 For each of the following numerical values of α, find $z_{\alpha/2}$:
   (a) $\alpha = .30$ (e) $\alpha = .015$
   (b) $\alpha = .15$ (f) $\alpha = .008$
   (c) $\alpha = .075$ (g) $\alpha = .002$
   (d) $\alpha = .03$ (h) $\alpha = .001$
3 If $z_\alpha = 1.41$, find α.
4 If $z_\alpha = 2.93$, find α.
5 If $z_{\alpha/2} = 2.19$, find α.
6 If $z_{\alpha/2} = 1.20$, find α.

SUMMARY AND DISCUSSION

In Chap. 2, we have continued our development of "descriptive statistics" by studying general patterns of data called probability distributions. We have concentrated especially on the binomial and normal distributions, the two most common types of distributions which describe behavior of data arising out of applications in the social, natural, and managerial sciences. The techniques developed in this chapter will be utilized throughout the remainder of the book as we enter the field of statistical inference, the drawing of conclusions (inferences) from a study of patterns inherent in a set of collected data. Here we have seen how to recognize some kinds of data which have the binomial distribution and other kinds which have the normal distribution. Future chapters will be devoted to special methods of analyzing binomial, normal, and other patterns of data. The notion of a probability distribution is a fundamental idea in the organized study of statistics, for we learn from it that sets of data are not merely haphazard collections of numbers but are, in fact, based upon general rules of behavior.

SUPPLEMENTARY EXERCISES

1   A company psychologist estimates that 30% of the employees of a large corporation are regular smokers of cigarettes. If her estimate is correct,
    (a) What is the probability that of 10 randomly selected employees, more than 4 are smokers?
    (b) What is the probability that of 100 randomly selected employees, more than 40 are smokers?

2   One public health physician claims that 60% of those with lung disorders are heavy smokers of cigarettes. If his assertion is correct,
    (a) Find the probability that of 20 such patients recently admitted to a hospital, fewer than half are heavy smokers.
    (b) Find the probability that of 200 such patients recently admitted, fewer than half are heavy smokers.

3   The latest figures indicate that 15% of all licensed drivers in one major metropolitan area have a drinking problem. If on one particular day 170 drivers apply for the normal three-year renewal of their licenses, what is the probability that more than 25% of them are problem drinkers?

4   One urban affairs sociologist feels that 55% of the adult residents of a particular major city have been victimized by a criminal at one time or another.
    (a) If his figures are correct, find the probability that from a random sample of 18 residents fewer than 5 have ever been victimized.
    (b) Under the sociologist's assumptions, what would be the probability that fewer than 500 of 1800 residents have ever been victimized?

5   An accountant reports to her superiors that 35% of all accounts which are currently overdue will eventually require legal action to force payment. If her estimate is correct,
    (a) What is the probability that of 12 currently overdue accounts fewer than 3 will actually require legal action?
    (b) Find the probability that of 120 overdue accounts fewer than 30 will require legal action.

6   A psychologist specializing in marriage counseling claims that his program

of "conscientious communication" can prevent divorce in 80% of the cases he works on.

(a) If his claims are true, what is the probability that of 20 couples currently involved in his program, more than 18 will stay together?

(b) What is the probability that of 200 cases in his files, fewer than 150 were able to avoid divorce?

7 The time it takes a college student to complete a standard psychological test is a normally distributed random variable with mean 80 minutes and standard deviation 10 minutes.

(a) How much time should be allowed for the test so that 90% of the students will be able to finish it?

(b) How much time should be allowed for the test so that 95% of the students will be able to finish it?

(c) How much time should be allowed for the test so that 99% of the students will be able to finish it?

8 A proposed test for the presence of diabetes assigns to each patient tested a number indicating that person's level of sugar in the urine. Patients with diabetes have levels which are normally distributed with mean 56 and standard deviation 3, while nondiabetic patients have levels which are normally distributed with mean 46 and standard deviation 10. The proposed test is intended to be primarily a preliminary indication of the presence of the disease, and so those patients having levels above 50.5 are scheduled for a more detailed battery of tests, while those with levels below 50.5 are considered to be nondiabetic.

(a) By the proposed test, what percentage of diabetic patients are incorrectly judged to be nondiabetic?

(b) What percentage of nondiabetic patients are unnecessarily scheduled for the more detailed battery of tests?

9 A particular job requires very relaxed individuals. A psychological test of aptitude is available for this job on which more relaxed persons score higher than less relaxed ones. The scores on this test are standardized to be normally distributed with mean 70 and standard deviation 10. What score is required to rank among the top 5% of all those taking the test?

10 The main plant of General Fabricators is lighted by several thousand light bulbs whose length of life is normally distributed with mean 2000 hours and standard deviation 100 hours. Instead of replacing each bulb as it burns out (individual attention requires greater expense overall), the plant replaces all bulbs at the same time on a regular schedule. The management would like to set up the replacement schedule in such a way that no more than 1% of the bulbs will have burned out before being replaced. After how many hours of use should all the bulbs be replaced in order to achieve this goal?

11 An economist studying the fluctuation of vegetable prices in small grocery stores in rural areas concludes that the weekly average of prices per pound of 10 basic vegetables has mean 30 cents and standard deviation 7.8 cents.

(a) If the economist does not know the probability distribution of the vegetable prices, what should his answer be to the question, "At most what proportion of the time does the weekly average exceed 48 cents per pound"?

(b) If the economist tests the data and finds the data to be normally distributed, how should he answer the question?

(c) Explain why the two answers are different, and explain the relationship between them.

12 A psychologist working for the National Safety Council is assigned the task of measuring the reaction times of experienced interstate drivers when they are confronted with an animal crossing the highway in front of them at night. She discovers that the reaction times are approximately normally distributed with mean 2.1 seconds and standard deviation .7 second. Earlier studies of animals crossing highways have found that if the driver does not react before 3.6 seconds have passed, the animal will be hit. What proportion of the time is the animal hit?

13 One mathematics instructor at a large university always gives A's to 10% of the students in his large introductory course, B's to 20%, C's to 40%, D's to 20% and F's to the remaining 10%. On one final exam, the scores were normally distributed with mean 70 and standard deviation 10. What ranges of scores would qualify students for A's, B's, C's, D's, and F's?

ESTIMATION

What we have been discussing so far in this course is generally referred to as "descriptive statistics." Descriptive statistics is that aspect of statistics whose primary objective is to describe, graphically and numerically, the characteristic behavior of sets of data. However, in most cases of applied interest, the data we have available will be only a small fraction of the total existing information on the subject; so we cannot be certain that the data we have will be a true representation of the existing situation. For example, suppose a distributor of citrus fruit wants to know the mean weight of California grapefruits for the purpose of estimating shipping costs. The way to calculate it would, of course, be to weigh each California grapefruit, add all the weights, and divide by the total number of grapefruits. To do this would be very difficult, if it were even possible. The best that can be done is to select a random sample of a relatively few grapefruits (a few compared to the total number of California grapefruits in existence) and calculate the mean weight of all the grapefruits in the sample. One of the most important questions in statistics then surfaces, "What is the relationship between this 'sample mean' and the figure we are looking for, the 'true mean' weight of all California grapefruits?" To answer this question requires the development of a second aspect of statistics, the aspect called "statistical inference." In statistical inference, our goal is to determine what conclusions about a situation may be validly and logically drawn (inferred) on the basis of a random sample of statistical data. The remainder of this book is primarily concerned with statistical inference.

SECTION 3.A
THE PROBLEM OF ESTIMATING THE TRUE MEAN

Let's denote the true mean weight of all California grapefruits by the Greek letter μ. How can we calculate μ? Well, we need to weigh all California grapefruits and then calculate the mean of all the weights. As a practical matter, however, it is virtually impossible to do this. Some of the reasons why the exact value of μ cannot be directly calculated are: (1) It would involve too much work and expense to remove every single mature California grapefruit from its tree and weigh it. (2) It would create an instantaneous glut on the grapefruit market exceeding the capacity of all cold storage facilities, thus wiping out a substantial portion of the citrus industry. (3) By the time the results were announced, they would be invalid because many of the grapefruits used in the study would have been eaten, while new ones (of possibly different weights) would have matured on the trees. (4) The results, even if obtainable, would be valid for only one season anyway, and it would be ridiculous to repeat this procedure every year. What we therefore need is an indirect way of getting a reasonably accurate measurement of μ, a way which is quite a bit simpler than the procedure just described.

Because it is not possible to weigh every single grapefruit, how many of them should we weigh? Well, it seems reasonable that the more grapefruits we weigh, the closer our average of those weighed will be to the true mean weight (if such a thing indeed exists) of all California grapefruits. Suppose, for example, we take a sample of n randomly selected California grapefruits. By "randomly selected," we mean chosen in such a way that every grapefruit in the state has an equal chance of winding up in the sample. From our random sample of n grapefruits, we compute the following numbers:

x_1 = weight of the 1st grapefruit of the sample

x_2 = weight of the 2d grapefruit of the sample

$\vdots$

x_k = weight of the kth grapefruit of the sample

$\vdots$

x_n = weight of the nth grapefruit of the sample

Using these n numbers, which are the weights of the n grapefruits in our random sample, we can compute the "sample mean"

★ $\quad \bar{x} = \dfrac{\Sigma x}{n} = \dfrac{x_1 + x_2 + \cdots + x_n}{n}$ $\qquad\qquad\qquad$ (3.1)

Here the sample mean, the mean of the numbers in the sample, is denoted by the symbol $\bar{x}$ (pronounced "x bar"). By way of contrast, μ is often referred to as the "population mean" to signify that it is the mean of an entire population, not merely of a sample.

Let's describe, if we can, the relationship between the sample mean $\bar{x}$ and the true mean μ. Does $\bar{x} = \mu$? If we choose n grapefruits at random and weigh them, will their mean weight be the same as the true mean weight of all California grapefruits? In answer to this question, we'd have to say "probably not," for it would be highly fortuitous if we were to pick n grapefruits (10, 500, 2000, or whatever number n may be) whose mean turned out to be exactly the same as the overall mean of all California grapefruits. In fact, suppose I pick n grapefruits, you pick n grapefruits, and six other people we know each pick n grapefruits. Then each of us will have his or her own personal $\bar{x}$. More likely than not, these eight $\bar{x}$'s not only will differ from μ, but they will also differ from each other. All eight of us will come up with a different value of $\bar{x}$. Of course, then, it is nonsensical to expect $\bar{x}$ to be the same as μ.

Now that we have agreed that $\bar{x}$ is probably not the same as μ, it would be useful to know how close it actually is to μ. The surprising answer to this question is that we really have no idea how close $\bar{x}$ is to μ. Our difficulty here is rooted in the fact that we don't really know the numerical value of μ; therefore, no matter what the value of $\bar{x}$ is, we can't tell how close it is to μ.

Well, the chances for an accurate estimate of μ look sort of bad, don't they? It looks as though we have gone just about as far as we can go. At this point, fortunately, we have available some major theoretical results from advanced mathematics. It is the job of professional mathematicians to discover ways of getting us out of situations like this, and their methods will be presented in the next section.

EXERCISES 3.A

1 Give at least one reason why each of the following true means cannot be calculated exactly:
   (a) The true mean protein content of eggs produced in Denmark.
   (b) The true mean height of American adults.
   (c) The true mean cost of weekly groceries for a family of four.
   (d) The true mean number of persons per household in the United States.

- (e) The true mean number of pupils in second-grade classrooms in the Chicago public school system.
- (f) The true mean ozone level in a cubic meter of air over Los Angeles County today.
- (g) The true mean rainfall in the state of Louisiana last week.
2 Explain why each of the following true means can be calculated exactly:
- (a) The true mean paid attendance per game at the home games of the New York Mets last season.
- (b) The true mean grade-point average of all students attending the University of Arizona last term.
- (c) The true mean number of votes cast per congressional district during the 1976 general election.
- (d) The true mean number of empty seats on all nonstop commercial flights between Boston and Chicago last month.

SECTION 3.B
CONFIDENCE INTERVALS FOR THE MEAN

Our objective in this section is to develop the procedures for estimating the true mean μ from a random sample. We will refer to random samples which contain more than 30 data points as "large samples." By "small samples," we mean sets of data consisting of 30 or fewer data points.

THE LAW OF AVERAGES

We noted earlier that, when we choose a sample of n data points and calculate their sample mean $\bar{x}$, we really don't know how close $\bar{x}$ is to the true mean μ. It seems reasonable, however, that the more data points we have in our sample, the closer $\bar{x}$ will be to μ. For example, if I tried to estimate the true mean weight of California grapefruits by randomly selecting 25 grapefruits and averaging their weights, while you tried to do the job with a sample of 200 grapefruits, we'd have to agree that your $\bar{x}$ would probably be closer to μ than my $\bar{x}$ would be. The formal statement of this principle is a mathematical theorem called the "law of large numbers," or more popularly, the "law of averages." A fairly simple wording of the law of large numbers reads as follows:

If a population has the true mean μ, and $\bar{x}$ is the sample mean of a random sample of n members of the population, then, as n grows larger and larger (as we take more and more data points for our sample), the sample mean $\bar{x}$ becomes closer and closer to the true mean μ.

Because of the law of large numbers, we know that we can improve our estimate of μ by using random samples of larger and larger amounts of data points. However, we unfortunately still do not know exactly how far we can expect our $\bar{x}$ to be from μ. And this, of course, is the key to finding out something about μ.

THE CENTRAL LIMIT THEOREM

A second mathematical theorem called the "central limit theorem" goes a long way toward breaking this impasse. Roughly speaking, it asserts that, for large values of n, the set of all possible sample means of random samples consisting of n data points is a set of normally distributed data. More precisely, we can state the central limit theorem as follows:

> If a population has the true mean μ and the true standard deviation σ, then the probability distribution of the set of all possible sample means of n members of the population becomes closer and closer to the normal distribution with mean μ and standard deviation $\sigma/\sqrt{n}$ as n grows larger and larger.

For all practical purposes, if n is larger than 30, namely, if we are working with a set of more than 30 data points, then the distribution of all possible sample means is close enough to the normal distribution for us to consider it to be the normal distribution. For large samples, then, we consider the numbers $\bar{x}$ to be normally distributed with mean μ and standard deviation $\sigma/\sqrt{n}$. This situation is illustrated in Fig. 3.1.

LARGE SAMPLES

In any particular large-sample applied problem under study, consequently, it is necessary to use Table A.3 of the Appendix, the table of the normal distribution.

FIGURE 3.1
Distribution of sample means for large samples.

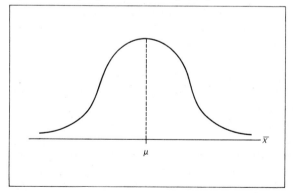

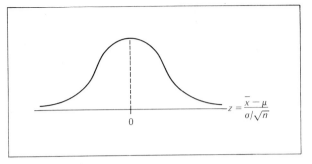

FIGURE 3.2
Distribution of z scores of sample means for large samples.

Now Table A.3, as you will recall, is actually a table of the z scores of the normal distribution, and it is therefore more useful to work with z scores of the possible sample means than with the original sample means themselves. A graph of the distribution of the z scores of the possible sample means, a standard normal distribution, appears in Fig. 3.2.

In terms of the percentage points of the normal distribution, this means that a fraction $1 - \alpha$, or a percentage $(1 - \alpha)100\%$, of the z scores fall between the numbers $-z_{\alpha/2}$ and $z_{\alpha/2}$. For example, taking $\alpha = .05$ so that $1 - \alpha = 1 - .05 = .95$ and $(1 - \alpha)100\% = .95 \times 100\% = 95\%$, we see that 95% of the z scores fall between -1.96 and 1.96. This situation is illustrated in Fig. 3.3. In probability language, we can express this fact by saying that the probability is 95% that for any particular sample mean $\bar{x}$ its z score

$$z = \frac{\bar{x} - \mu}{\sigma/\sqrt{n}}$$

will fall between -1.96 and 1.96.

FIGURE 3.3
The middle 95% of z scores of sample means (large samples).

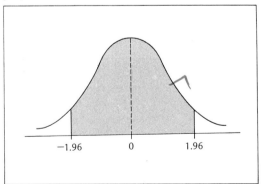

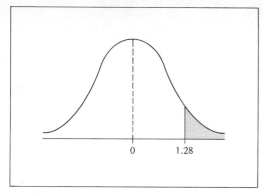

FIGURE 3.4
The upper 10% of z scores of sample means (large samples).

Figure 3.4 provides another view of how the central limit theorem works. That graph shows that 10% of the z scores fall above 1.28. We can therefore say that 10% of the sample means result in a z score above 1.28, so that there is a 10% chance of your z score (or mine or anybody else's, for that matter) will turn out to be larger than 1.28.

Unfortunately, although we can be fairly sure that 10% of the z scores will be larger than 1.28, we can never tell whether a particular z score is among that upper 10%. Why is this so? Well, the reason goes back to the fact that we do not know what μ is, and therefore we can never actually calculate the value of the z score

$$z = \frac{\bar{x} - \mu}{\sigma/\sqrt{n}}$$

In fact, not only do we not know what μ is, but we also do not generally know what σ is. Therefore we cannot substitute enough numbers into the formula for z in order to calculate it. However, even though we cannot tell whether a particular z score is, for example, between −1.96 and 1.96, we do know that 95% of them are. And we can use this information to help us in estimating the numerical value of the true mean μ.

We first express in mathematical symbols the statement that 95% of the z scores fall between −1.96 and 1.96. In algebraic language,[1] we can be 95% sure that

$$-1.96 < z < 1.96$$

---

[1]The notation $-1.96 < z < 1.96$ represents a combination of the two statements $-1.96 < z$ (z is greater than −1.96) and $z < 1.96$ (z is less than 1.96).

Replacing z by its formula, we see that we can be 95% sure that

$$-1.96 < \frac{\bar{x} - \mu}{\sigma/\sqrt{n}} < 1.96$$

Multiplying across the above inequality (i.e., an equation with $<$ or $>$ instead of $=$) by the denominator $\sigma/\sqrt{n}$ gives us 95% certainty that

$$-1.96\frac{\sigma}{\sqrt{n}} < \bar{x} - \mu < 1.96\frac{\sigma}{\sqrt{n}}$$

Applying a final bit of algebra, we see that we can be 95% sure that

$$\bar{x} - 1.96\frac{\sigma}{\sqrt{n}} < \mu < \bar{x} + 1.96\frac{\sigma}{\sqrt{n}}$$

Now is a good time to go back to the English language and reflect upon what the resulting statement really means. A direct translation of the last mathematical statement just above would read as follows: We can be 95% sure that the true mean μ falls between $\bar{x} - 1.96\ (\sigma/\sqrt{n})$ and $\bar{x} + 1.96\ (\sigma/\sqrt{n})$,

where $\bar{x}$ = sample mean based on our particular random sample of n members of the population

σ = true standard deviation of the population

n = number of data points in our sample

Let's return to the problem introduced in Sec. 3.A, the problem of finding the true mean weight of California grapefruits. Suppose we try to estimate the true mean μ by selecting a sample of $n = 64$ grapefruits. We weigh each of these grapefruits, and they turn out to have a mean weight of $\bar{x} = 11.36$ ounces. What we now want to know is how close we can expect our sample of 11.36 to be to the true mean μ. Well, by the development above, we can be 95% sure that

$$11.36 - 1.96\frac{\sigma}{\sqrt{64}} < \mu < 11.36 + 1.96\frac{\sigma}{\sqrt{64}}$$

Taking note of the fact that $\sqrt{64} = 8$ and that $1.96/8 = .25$, it follows that we can be 95% sure that the true mean weight μ of California grapefruits falls between $11.36 - .25\sigma$ ounces and $11.36 + .25\sigma$ ounces, where σ is the true standard deviation of the weights of all California grapefruits. Until we find out the numerical value of σ, we therefore will not know how good our estimate of μ is. Unfortunately, the same difficulties that prevented us from directly measuring μ also prevent us from directly measuring σ.

It turns out that for large values of n (i.e., $n > 30$) there is a way of estimating σ to a degree of accuracy sufficient for completing the problem at hand. What we do is replace σ in the formula we have been using by a number s, called the "sample standard deviation," which is an estimate of the true standard deviation calculated from the n data points of the random sample. The formula for the sample standard deviation s is a slight variation of the formula for σ introduced in Sec. 1.D. It is as follows:

$$\bigstar \quad s = \sqrt{\frac{\Sigma(x - \bar{x})^2}{n - 1}} \qquad\qquad (3.2)$$

Comparing this latter formula with the one for σ in Chap. 1, we note two changes: (1) μ is replaced by $\bar{x}$. (2) In the denominator, n is replaced by $n - 1$. The change for μ to $\bar{x}$ is necessary because we are dealing with a random sample of which we know (or can calculate) the sample mean, but we have no access to the true mean μ. The replacement of n by $n - 1$ is less easy to understand. Basically, the idea here is that if we calculate s according to the formula presented above, the mean of all possible values of s obtainable from samples of size n will be σ, in the same way that the mean of all values of $\bar{x}$ is μ. If we use n, however, instead of $n - 1$ in the denominator of s, the possible values of s will average out to something slightly different than σ.

As an equivalent shortcut formula for the sample standard deviation, we have

$$\bigstar \quad s = \sqrt{\frac{n\Sigma x^2 - (\Sigma x)^2}{n(n - 1)}} \qquad\qquad (3.3)$$

Suppose our random sample of 64 grapefruits turned out to have sample standard deviation $s = 1.76$ ounces. Replacing σ in the expression

$$11.36 - .25\sigma < \mu < 11.36 + .25\sigma$$

by $s = 1.76$ implies that we can be 95% sure that, successively,

$$11.36 - (.25)(1.76) < \mu < 11.36 + (.25)(1.76)$$
$$11.36 - .44 < \mu < 11.36 + .44$$
$$10.92 < \mu < 11.80$$

Our conclusion about the mean weight of California grapefruits can now be expressed as follows: On the basis of our random sample of 64 California grapefruits, we can be 95% sure that the true mean weight of all California grapefruits is somewhere between 10.92 and 11.80 ounces.

The interval 10.92 to 11.80 is said to be a 95% "confidence interval" for μ. It allows us to come up with not only an estimate of μ but also a measure of how accurate we can expect our estimate to be. A picture of this confidence interval appears in Fig. 3.5.

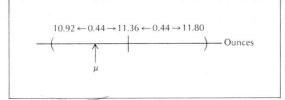

10.92 ← 0.44 → 11.36 ← 0.44 → 11.80 — Ounces

μ

FIGURE 3.5
A 95% confidence interval for the mean weight of California grapefruits.

The confidence interval we have just found is only one of many, many confidence intervals that may be used to estimate μ. In fact, it is not *the* confidence interval for μ, but is is *a* 95% confidence interval for μ based on *a* random sample of size 64. If, for example, we wanted to be 98% sure of having an interval in which μ fell, we would have to modify our present interval somewhat, by replacing the 1.96 in the formula (and in Fig. 3.3) by $z_{.01} = 2.33$. Based on the same sample of 64 data points, we could be 98% sure that

$$11.36 - (2.33)\frac{1.76}{\sqrt{64}} < \mu < 11.36 + (2.33)\frac{1.76}{\sqrt{64}}$$

namely, that $10.85 < \mu < 11.87$. This says that the mean weight of California grapefruits lies somewhere between 10.85 and 11.87 ounces.

Notice that the 98% confidence interval for μ is wider than the 95% interval based on the same data. It is logical that this be so, because if we want to be more certain of having an interval which contains μ, we are going to have to give ourselves a little more room by widening our interval. Similarly, if we were to narrow our interval, there might be only an 80% chance of its containing μ. We summarize this discussion in Table 3.1, which demonstrates the procedure of obtaining confidence intervals of various degrees of confidence. For purposes

TABLE 3.1
Various confidence intervals for the true mean weight of California grapefruits
(Based on a sample of $n = 64$ grapefruits having sample mean $\bar{x} = 11.36$ and sample standard deviation $s = 1.76$)

Degree of Confidence $(1 - \alpha)100\%$	α	$\alpha/2$	$z_{\alpha/2}$	$z_{\alpha/2}\frac{s}{\sqrt{n}}$	Confidence Interval $\bar{x} - z_{\alpha/2}\frac{s}{\sqrt{n}} < \mu < \bar{x} + z_{\alpha/2}\frac{s}{\sqrt{n}}$
80%	.20	.10	1.28	.28	$11.08 < \mu < 11.64$
90%	.10	.05	1.64	.36	$11.00 < \mu < 11.72$
95%	.05	.025	1.96	.44	$10.92 < \mu < 11.80$
98%	.02	.01	2.33	.51	$10.85 < \mu < 11.87$
99%	.01	.005	2.57	.57	$10.79 < \mu < 11.93$
99.9%	.001	.0005	3.3	.73	$10.63 < \mu < 12.09$

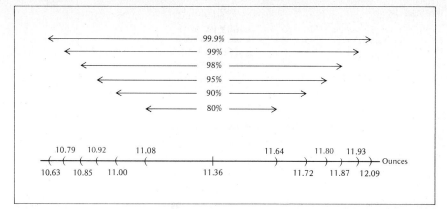

FIGURE 3.6
Graphical representations of various confidence intervals for the true mean weight of California grapefruits.

of comparison the various intervals are graphed in Fig. 3.6. The calculations therein are based on the fact (discussed in detail in Sec. 2.E) that the middle $(1 - \alpha)$ 100% of the standard normal distribution lies between the numbers $-z_{\alpha/2}$ and $z_{\alpha/2}$. This fact results in the replacement of the 1.96 in our discussion by $z_{\alpha/2}$. We can therefore be $(1 - \alpha)$ 100% sure that

$$\bigstar \quad \bar{x} - z_{\alpha/2}\frac{s}{\sqrt{n}} < \mu < \bar{x} + z_{\alpha/2}\frac{s}{\sqrt{n}} \qquad (3.4)$$

based on a random sample of n data points having sample mean $\bar{x}$ and sample standard deviation s. Equation (3.4) is the general formula for computing confidence intervals for means based on large samples.

Example 3.1 Automobile Usage A random sample of 100 private automobiles registered in the province of Ontario put a mean yearly mileage of 6500 on their odometers with a standard deviation of 1400 miles. These facts were brought out as part of a statistical study of automobile usage commissioned by a local insurance agency. Find a 99% confidence interval for the true mean yearly mileage of all private cars in the province.

SOLUTION   For a 99% confidence interval, we have that

$$(1 - \alpha)100\% = 99\%$$

Dividing through by 100%, we get

$$1 - \alpha = .99$$

so that

$$-\alpha = .99 - 1 = -.01$$

and finally

$$\alpha = .01$$

It follows that

$$\frac{\alpha}{2} = .005$$

from which we conclude that

$$z_{\alpha/2} = z_{.005} = 2.57$$

The data gives us the information that

$$n = 100$$

$$\bar{x} = 6500$$

$$s = 1400$$

Applying formula (3.4), we know that

$$\bar{x} - z_{\alpha/2}\frac{s}{\sqrt{n}} < \mu < \bar{x} + z_{\alpha/2}\frac{s}{\sqrt{n}}$$

where μ is the true mean yearly mileage of all private cars in the province. Inserting the actual numerical values of the quantities involved yields the following calculations:

$$6500 - (2.57)\frac{1400}{\sqrt{100}} < \mu < 6500 + (2.57)\frac{1400}{\sqrt{100}}$$

$$6500 - \frac{(2.57)(1400)}{10} < \mu < 6500 + \frac{(2.57)(1400)}{10}$$

$$6500 - 360 < \mu < 6500 + 360$$

$$6140 < \mu < 6860$$

Therefore we can be 99% sure that the true mean yearly mileage of all private cars in the province of Ontario lies between 6140 and 6860 miles.

SMALL SAMPLES

Now that we have shown how to estimate the true mean of a population if we have a random sample of more than 30 data points, let's look at the small-sample situation. Sometimes it is not possible or economically feasible to use a large sample. For example, a steel mill may want to estimate temperatures, pressures, etc., in various parts of its furnaces and may not have as many as 30 furnaces available for testing. A political scientist may want to study the voting patterns of a particular occupational group, but data may not be available for 30 elections in which this group has participated. In any situations such as these, where sufficient data is unobtainable, the central limit theorem simply does not apply. We therefore have no guarantee that the z scores of the sample means are normally distributed. Use of the normal distribution in establishing confidence intervals then cannot be logically justified, and so a new approach to the problem is needed.

As it turns out, we can handle the case of small samples (30 or fewer data points) if we have reason to believe that the data points came from a normally distributed population. This will be the case, for example, in situations satisfying those criteria for normally distributed data presented at the beginning of Sec. 2.C. Just as the central limit theorem revealed (in the large-sample case) that the numbers

$$z = \frac{\bar{x} - \mu}{\sigma/\sqrt{n}}$$

are approximately normally distributed, mathematical calculations show that if the overall population from which a random sample is selected has the normal distribution, then for all samples of size n (no matter how large or small), the numbers

$$t = \frac{\bar{x} - \mu}{s/\sqrt{n}}$$

have the t distribution of Table A.4 of the Appendix

where $\bar{x}$ = sample mean
s = sample standard deviation
n = number of data points in the sample
μ = true mean of the population

The use of Table A.4 is valid only when the overall population has the normal distribution, because Table A.4, the table of percentage points of the t distribution, is derived by mathematical calculations from Table A.3, the table of the normal distribution.

Even though the numbers

$$t = \frac{\bar{x} - \mu}{s/\sqrt{n}}$$

have the t distribution for all values of n, both large and small, Table A.4 is commonly used only for small samples of $n \leq 30$. The reason for this is that for large values of n the t distribution is very close to the normal distribution due to the increasing influence of the central limit theorem. Common statistical practice therefore calls for the use of Table A.4 when $n \leq 30$ and Table A.3 when $n > 30$.

Beyond the switch from Table A.3 to Table A.4, there is no recognizable difference in the calculation of confidence intervals. In formula (3.4), we simply replace $z_{\alpha/2}$ by $t_{\alpha/2}[n - 1]$, the appropriate number selected from Table A.4. Here the number $n - 1$ is called the "degrees of freedom" and appears in the column headed df in the table. It is through the degrees of freedom that the number of data points makes its influence felt on the t distribution. We find the proper t value at the intersection of the row corresponding to the number $n - 1$ (one less than the number of data points) and the column headed $t_{\alpha/2}$ (for the relevant value of α). The formula for confidence intervals based on small samples is therefore

$$\star \quad \bar{x} - t_{\alpha/2}[n - 1]\frac{s}{\sqrt{n}} < \mu < x + t_{\alpha/2}[n - 1]\frac{s}{\sqrt{n}} \qquad (3.5)$$

The following two examples illustrate the use of this formula, particularly how to choose the correct value of $t_{\alpha/2}[n - 1]$.

Example 3.2 Company Cars A corporation which maintains a large fleet of company cars for the use of its sales staff needs to determine the average number of miles driven monthly per salesperson. A random sample of 16 monthly car-use records was examined, yielding the information that the 16 salespersons under study had driven a mean of 2250 miles with a standard deviation of 420 miles. Find a 95% confidence interval for the true mean monthly number of miles per salesperson for the entire sales staff.

SOLUTION   We can summarize the information contained in the sample by recording that

$n = 16$

$\bar{x} = 2250$

$s = 420$

TABLE 3.2
A portion of Table A.4

df	$t_{.025}$	df
15	2.131	15

Because $n = 16 < 30$, use of the percentage points of the t distribution, rather than those of the normal distribution, is required. Since we want a 95% confidence interval, we know that $\alpha = .05$ so that $\alpha/2 = .025$. As $n = 16$, we know that df $= n - 1 = 16 - 1 = 15$; so the number we need from Table A.4 is $t_{\alpha/2}[n - 1] = t_{.025}[15]$. We reproduce the relevant portion of Table A.4 in Table 3.2, which shows that $t_{.025}[15] = 2.131$. Now we have in our possession all the components of formula (3.5) for confidence intervals based on small samples, where μ stands for the true mean monthly number of miles per salesperson for the entire sales staff. Therefore, inserting the numerical values for n, x, s, and $t_{\alpha/2}[n - 1]$, we see that we can be 95% sure that

$$2250 - (2.131)\frac{420}{\sqrt{16}} < \mu < 2250 + (2.131)\frac{420}{\sqrt{16}}$$

$$2250 - \frac{(2.131)(420)}{4} < \mu < 2250 + \frac{(2.131)(420)}{4}$$

$$2250 - 223.8 < \mu < 2250 + 223.8$$

and finally

$$2026.2 < \mu < 2473.8$$

Therefore, we can be 95% sure that the true mean monthly number of miles driven per salesperson for the entire sales staff is somewhere between 2026.2 and 2473.8 miles.

Example 3.3 Manufacture of Antibiotics Because pharmaceutical companies manufacture antibiotics in large vats for later bottling in smaller units, it is necessary to test potency levels at several locations in the vats before bottling. While it is impossible to achieve complete uniformity in potency levels of the bottled product, such testing is required for the purpose of obtaining the mean potency level of the entire batch. In one such test, the readings at 12 randomly selected locations in the vat were

8.9	9.0	9.1	8.9	9.1	9.0
9.0	9.0	8.9	8.8	9.1	9.2

TABLE 3.3
Calculation of x and s for the antibiotic potency data

x	$x - \bar{x}$	$(x - \bar{x})^2$
8.9	−.1	.01
9.0	.0	.00
9.1	.1	.01
8.9	−.1	.01
9.1	.1	.01
9.0	.0	.00
9.0	.0	.00
9.0	.0	.00
8.9	−.1	.01
8.8	−.2	.04
9.1	.1	.01
9.2	.2	.04
108.0	.0	.14

$$\bar{x} = \frac{108.0}{12} = 9.0$$

$$s = \sqrt{\frac{.14}{11}} = \sqrt{.0127} = .113$$

Find a 98% confidence interval for the true mean potency reading for the entire batch.

SOLUTION   In this problem, we know immediately that $n = 12$, but we will have to compute $\bar{x}$ and s from the data. We proceed to make the calculations by the methods introduced in Chap. 1. The basic table of calculations appears in Table 3.3 and shows that $\bar{x} = 9$ and $s = .113$. Because we are looking for a 98% confidence interval, we are using $\alpha = .02$ so that $\alpha/2 = .01$. For $n = 12$, we know that df $= n - 1 = 12 - 1 = 11$, and so we obtain from Table A.4 the relevant value of $t_{\alpha/2}[n - 1]$; namely,

$$t_{.01}[11] = 2.718$$

as shown in Table 3.4. We insert these numbers into formula (3.5) for confidence intervals based on small samples, where μ is the true mean potency level of the entire batch. On the basis of the data collected, we can then be 98% sure that

$$9.0 - (2.718)\frac{.113}{\sqrt{12}} < \mu < 9.0 + (2.718)\frac{.113}{\sqrt{12}}$$

$$9.0 - \frac{(2.718)(.113)}{3.464} < \mu < 9.0 + \frac{(2.718)(.113)}{3.464}$$

$$9.0 - .09 < \mu < 9.0 + .09$$

TABLE 3.4
A portion of Table A.4

df	$t_{.01}$	df
11	2.718	11

and finally

$$8.91 < \mu < 9.09$$

We can therefore be 98% sure that the true mean potency level of the entire batch of antibiotics in the vat falls between 8.91 and 9.09.

EXERCISES 3.B

1   For 64 randomly selected months in the past, the Oryx Veterinary Supply Company (OVS) has sold an average of 16,224.72 dollars worth of goods each month with a standard deviation of 205 dollars. Find a 90% confidence interval for the true mean monthly sales level of OVS.

2   A biologist studying the water life of the Gulf of California has found 36 examples of a certain breed of fish. The 36 fish have a mean length of 11 inches with a standard deviation of 1.5 inches. What can she present as a 95% confidence interval for the true mean length of all fish of that breed living in the Gulf of California?

3   An auto-parts dealer plans on advertising the average lifetime of the auto batteries he sells. To find out what the average lifetime is, he gives away 100 randomly selected batteries to 100 randomly selected customers, with the proviso that the customer use the battery continuously with no maintenance except for adding water as needed and cleaning the terminals until it fails. The lifetimes of the 100 batteries turn out to have mean 2.0 years with a standard deviation of .156 years.

   (a) Determine a 90% confidence interval for the true mean lifetime of the dealer's batteries.

   (b) Determine a 95% confidence interval for the true mean lifetime.

   (c) Determine a 99% confidence interval for the true mean lifetime.

4   For 18 randomly selected months in the past, the Oryx Veterinary Supply Company (OVS) has sold an average of 16,224.72 dollars worth of goods each month with a standard deviation of 205 dollars. Find a 90% confidence interval for the true mean monthly sales level of OVS.

5   The Desert Breeze Casualty Insurance Company of Trona wants to estimate the mean number of auto accidents its clients cause per month. They have collected the following data over 12 randomly selected months in the recent past:

Month	No. of Accidents
May 1974	24
July 1974	16
October 1974	20
November 1974	18
January 1975	22
March 1975	8
June 1975	12
July 1975	28
September 1975	20
December 1975	14
April 1976	10
October 1976	12

Find an 80% confidence interval for the true mean number of auto accidents caused per month by the company's clients.

6 An auto-parts dealer plans on advertising the average lifetime of the auto batteries he sells. To find out what the average lifetime is, he gives away 10 randomly selected batteries to 10 randomly selected customers, with the proviso that the customer use the battery continuously with no maintenance except for adding water as needed and cleaning the terminals until it fails. The lifetimes, in years, of the 10 batteries turn out to be as follows:

   2.2   1.9   2.0   2.1   1.8   2.1   2.1   2.1   1.7   2.0

   (a) Determine a 90% confidence interval for the true mean lifetime of the dealer's batteries.
   (b) Determine a 95% confidence interval for the true mean lifetime.
   (c) Determine a 99% confidence interval for the true mean lifetime.

7 A random sample of 25 cigarettes of the Puffer's Heaven brand had a mean nicotine content of 22 milligrams with a standard deviation of 4 milligrams. Find an 80% confidence interval for the true mean nicotine content of Puffer's Heaven cigarettes.

8 A group of fifth-grade children who have participated in a special remedial reading program are tested for reading improvement. A random sample of 81 children had a mean increase of 18.4 points with a standard deviation of 6.9. Find a 97% confidence interval for the mean increase in reading scores of the entire group.

SECTION 3.C
ESTIMATING PROPORTIONS

The objective of a political poll is to estimate the proportion of registered voters favoring one or another candidate or position. The objective of a test of effectiveness of a new drug is to estimate the proportion of cases satisfactorily treated by the drug. And the objective of scholastic examinations is to estimate the proportion of subject matter learned by each student.

When we are estimating a proportion, we collect a set of binomially distributed data having parameters n (the number of data points) and p (the true proportion we want to estimate). We can therefore use the facts about the binomial distribution, recorded in the last paragraph of Sec. 2.B, that

$$\mu = np$$

and

$$\sigma = \sqrt{np(1 - p)}$$

We can insert these expressions into formula (3.4) for large-sample confidence intervals, replacing p by $\hat{p}$ (pronounced "p hat"), the "sample proportion" based on the data. (For example, $\hat{p}$ could be the proportion of sampled voters who favor the position in question, the proportion of test cases satisfactorily treated by the new drug, or the proportion of examination questions correctly answered by the student.) Then formula (3.4) leads to the $(1 - \alpha)$ 100% confidence interval formula for the true proportion p

$$\star \quad \hat{p} - z_{\alpha/2}\sqrt{\frac{\hat{p}(1 - \hat{p})}{n}} < p < \hat{p} + z_{\alpha/2}\sqrt{\frac{\hat{p}(1 - \hat{p})}{n}} \tag{3.6}$$

The above formula is valid for $n > 20$, essentially the domain of validity of the normal approximation to the binomial distribution.

Example 3.4 A Political Poll (This example continues the framework of Examples 2.2 and 2.8.) Of a sample of 700 voters, 390 favor repeal of a certain local tax law. Find a 95% confidence interval for the true proportion of all voters favoring repeal.

SOLUTION   Here we have $\alpha = .05$ so that $z_{\alpha/2} = 1.96$. We have $n = 700$, and we see that the sample proportion favoring repeal is

$$\hat{p} = \frac{390}{700} = .557$$

Therefore, denoting by p the true proportion favoring repeal, we can be 95% sure, in view of formula (3.6), that, successively,

$$.557 - 1.96\sqrt{\frac{.557(1 - .557)}{700}} < p < .557 + 1.96\sqrt{\frac{.557(1 - .557)}{700}}$$

$$.557 - 1.96\sqrt{\frac{(.557)(.443)}{700}} < p < .557 + 1.96\sqrt{\frac{(.557)(.443)}{700}}$$

$$.557 - 1.96\sqrt{.000353} < p < .557 + 1.96\sqrt{.000353}$$

$$.557 - (1.96)(.0188) < p < .557 + (1.96)(.0188)$$

and finally

$$.557 - .037 < p < .557 + .037$$

We conclude that, on the basis of the 700 voters sampled, we can be 95% sure that the true proportion favoring repeal is somewhere between .520 and .594, namely, between 52% and 59.4%. (In particular, we are over 95% sure that a majority of voters favors repeal.)

EXERCISES 3.C

1 The manufacturer of a new chemical fertilizer would like to know the proportion of seeds that will germinate if they are planted in soil containing the fertilizer. On the basis of an experiment in which 130 seeds germinated out of 170 that were planted, find a 95% confidence interval for the true proportion of such seeds that will germinate in soil containing the fertilizer.

2 An insurance company is planning to offer a 10-year term group life insurance policy to all 45-year-old men in a certain occupational grouping. In order to determine a reasonable premium, the company conducts an actuarial study of all men of that age in that occupational grouping. If a survey of 2000 such men in the past showed that 150 died before reaching age 55, construct a 99% confidence interval for the true proportion of such men who die before age 55.

3 A news reporter would like to predict the percentage of members of the 50 state legislatures who favor a certain controversial change in the banking laws. If a poll of 120 legislators yields 72 who support the change, find a 90% confidence interval for the true percentage favoring the change.

4 A company psychologist would like to estimate the proportion of regular smokers of cigarettes among the company's employees. Research turns up 45 smokers in a random sample of 195 employees.
   (a) Find an 80% confidence interval for the true proportion of regular smokers among the company's employees.
   (b) Find a 98% confidence interval for the true proportion.

5 A sociologist studying the prevalence of crime in one major city asks 1800 randomly selected residents whether or not they have ever been victimized by a criminal. Of those surveyed, 700 responded in the affirmative.

(a) Find a 90% confidence interval for the true percentage of city residents who have been victimized.

(b) Find a 95% confidence interval for that percentage.

(c) Find a 99% confidence interval for the percentage.

SECTION 3.D
CHOOSING THE SAMPLE SIZE

When we made the statement in Sec. 3.B that we can be 95% sure that the true mean μ falls between $11.36 - .44$ and $11.36 + .44$, we were saying that we are 95% sure that μ is within a distance .44 of 11.36. This way of expressing the situation can be seen in Fig. 3.5, where attention is drawn to the fact that 10.92 is located a distance .44 to the left of 11.36, while 11.80 is .44 to the right. Therefore, if we were to use the sample mean of 11.36 ounces as an estimate of the true mean μ, we could be 95% sure of being in error by no more than .44 ounce. If we look at the general situation this way, we can define the "error of estimation"

$$\bigstar \qquad E = z_{\alpha/2}\frac{s}{\sqrt{n}} \qquad\qquad (3.7)$$

in the following terms: We can be $(1 - \alpha)100\%$ sure that the maximum possible error in using $\bar{x}$ as an estimate of μ cannot exceed E. The two procedures, confidence interval and error of estimation, are merely ways of viewing the same information from different vantage points.

A glance at the error formula (3.7) reveals the following interesting[1] fact: As n increases, E decreases. In other words, the more data points we use in our sample, the smaller our error of estimation will be. For example, we have already seen that, for $n = 64$, we get $E = .44$. If $n = 144$, we will have $E = z_{\alpha/2}s/\sqrt{n} = (1.96)(1.76)/\sqrt{144} = .29$, and if we use a sample of $n = 225$ data points, our error will be $E = (1.96)(1.76)/\sqrt{225} = .23$. Table 3.5 contains a list of sample sizes together with their corresponding errors of estimation, so that we can trace a typical relationship between E and n.

Because of the relationship between sample size and error of estimation, as expressed by formula (3.7), we can even do better than knowing how accurate our estimates are. We can decide in advance how accurate we want to be, and then formula (3.7) will tell us how large a sample we need in order to achieve the desired level of accuracy. The following two examples illustrate how this process works.

Example 3.5. Soft-Drink Machines Owners of a chain of coin-operated beverage machines advertise that their machines dispense, on the average, 7 ounces

---

[1] I'll let you be the judge of this!

TABLE 3.5
Relationship between sample size
and error of estimation
(Based on a 95% confidence interval for the true
mean of a population having standard deviation 1.76)

n	$z_{\alpha/2}$	s	$E = z_{\alpha/2}\dfrac{s}{\sqrt{n}}$
36	1.96	1.76	.57
64	1.96	1.76	.44
100	1.96	1.76	.34
144	1.96	1.76	.29
196	1.96	1.76	.25
225	1.96	1.76	.23
400	1.96	1.76	.17
900	1.96	1.76	.11
2500	1.96	1.76	.07

of soft drink per cup. A consumer-testing organization is interested in the question of whether a particular machine is meeting its specifications. It wants to run a study after which it can be 95% sure of having the true mean amount dispensed by the machine correctly estimated to within an error of .01 ounce (one-hundredth of an ounce). If a preliminary analysis indicates that the amounts dispensed have standard deviation .12 ounce (twelve-hundredths of an ounce), how many sample cupfuls does the organization need to measure in order to achieve the desired level of accuracy?

SOLUTION   The unknown here is n, the number of data points required for the analysis, in this case, the number of sample cupfuls needed. The acceptable level of error is $E = .01$, and the desired degree of confidence is 95%. Therefore $\alpha = .05$ and $\alpha/2 = .025$, so that $z_{\alpha/2} = z_{.025} = 1.96$.
　　Finally, in place of the sample standard deviation (which is unavailable, for we have not yet taken the actual sample), we use the preliminary estimate that $s = .12$. Inserting these numbers into their proper places in the error formula (3.7), we have

$$.01 = (1.96)\frac{.12}{\sqrt{n}}$$

Successive steps of algebraic operations yield

$$.01\sqrt{n} = (1.96)(.12)$$

$$\sqrt{n} = \frac{(1.96)(.12)}{.01}$$

$$\sqrt{n} = 23.52$$

It then follows that $n = (23.52)^2 = 553.19$. To achieve the desired level of accuracy, then, would require at least 553.19 data points. Therefore, we would need 554 sample cupfuls for the survey. (Notice that even though 553.19 rounds down to 553, the latter number would leave us just short of the desired accuracy. That 554th data point puts us over the top.)

Using the error formula (3.7), we can derive an explicit formula for the required sample size n. Multiplying expression (3.7) through by $\sqrt{n}$ gives us

$$E\sqrt{n} = z_{\alpha/2}s$$

and dividing by E yields

$$\sqrt{n} = \frac{z_{\alpha/2}s}{E}$$

If we square both sides, we get the "sample size formula" for estimating means

$$\star \quad n = \left(\frac{z_{\alpha/2}s}{E}\right)^2 \tag{3.8}$$

In Example 3.5, we have $z_{\alpha/2} = z_{.025} = 1.96$, $s = .12$, and $E = .01$ and so we would have

$$n = \left[\frac{(1.96)(.12)}{.01}\right]^2 = (23.52)^2 = 553.19$$

using formula (3.8). This is, of course, the same result we obtained earlier.

We can also use the confidence interval formula (3.6) for proportions in order to determine how many data points are needed to achieve a desired level of accuracy. By analogy with the technique developed for means, we start with the error formula

$$E = z_{\alpha/2}\sqrt{\frac{\hat{p}(1 - \hat{p})}{n}}$$

where $\hat{p} - E < p < \hat{p} + E$. We solve the error formula for n to get

$$n = \left[\frac{z_{\alpha/2}\sqrt{\hat{p}(1 - \hat{p})}}{E}\right]^2$$

As we have not yet taken the data (because we are trying to figure out how many data points to use), we do not know what the value of $\hat{p}$ will be, and so we really cannot use the above formula for n. However, it is a fact of arithmetic that $\hat{p}(1 - \hat{p})$ will *always* be less than .25, no matter what number (between 0 and 1, of course) $\hat{p}$ turns out to be.[1] Therefore $\sqrt{\hat{p}(1 - \hat{p})}$ will always be less than $\sqrt{.25} = .5$, so that no matter what the eventual value of $\hat{p}$ turns out to be, it will always be sufficient to use

percentage or proportion formula

$$\bigstar \quad n = \left[\frac{z_{\alpha/2}(.5)}{E}\right]^2 \qquad\qquad (3.9)$$

data points.

Example 3.6 Effectiveness of a New Drug (This example is related to Example 2.3.) A new drug for alleviating mental depression is proposed, and we want to estimate its effectiveness. If we want to be 98% sure of having the true proportion of patients helped by the drug correctly estimated to within .03, how many patients do we need for our study?

SOLUTION Here $\alpha = .02$ so that $z_{\alpha/2} = z_{.01} = 2.33$. Furthermore, the desired level of accuracy is given by $E = .03$. Using formula (3.9) we see that a sufficient number of data points for our study would be

$$n = \left[\frac{z_{\alpha/2}(.5)}{E}\right]^2 = \left[\frac{(2.33)(.5)}{.03}\right]^2 = (38.83)^2$$

$$= 1508$$

Therefore a sample of 1508 patients will be sufficient to guarantee the level of accuracy desired in our study of the drug's effectiveness. On the basis of a sample of that size we can be 98% sure of having the true proportion of patients helped correctly estimated to within .03.

As an aside to the conclusion of the above example, it is interesting to note that the Gallup Poll, an organization which undertakes to predict the outcome of elections, uses a sample of 1500 voters in its surveys.

EXERCISES 3.D

1 An agricultural zoologist wants to be 95% sure of having the true mean gestation period of a new variant of dairy cow estimated correctly to within 1.3 days. It seems reasonable to use a standard deviation of 5 days for the variant under study. On how many cows are data needed in order to achieve the desired level of accuracy?
2 The Eastingfield Electric Company, which produces light bulbs, is required by law to estimate statistically the true mean lifetime of its bulbs. In particu-

---

[1]Try a few numerical values of $\hat{p}$ to confirm this fact.

lar, it has to be 95% sure of having the true mean estimated accurately to within 25 hours. If the lifetimes have standard deviation 200 hours, how many bulbs must be tested in order to achieve the desired accuracy in estimating the true mean?

3 A manufacturer of a new chemical fertilizer would like to estimate the proportion of seeds that will germinate if they are planted in soil containing the new fertilizer. If the manufacturer wants to be 95% sure of having the true proportion correctly estimated to within .02, how many seeds are required for the experiment?

4 In the insurance situation described in Exercise 3.C.2, records on how many men are required in order to be 99% sure of having the true percentage (of 45-year-old men in the occupational grouping who die before age 55) correctly estimated to within 1%?

5 How many residents must the sociologist of Exercise 3.C.5 poll in order to be 95% sure of having the true percentage of those victimized by a criminal estimated correctly to within 1.5%?

SUMMARY AND DISCUSSION

This has been the first chapter dealing with statistical inference, the science of drawing conclusions from information contained in a sample of data. The major problem of statistical inference is the attempt to determine to what extent a sample can be considered to be representative of the underlying population. Using two spectacular mathematical results, the law of large numbers and the central limit theorem, we have developed the method of confidence intervals to solve that problem. We first presented confidence interval formulas which show the extent to which the sample mean represents the true population mean and the sample proportion represents the true population proportion. We concluded by exhibiting a method of calculating how large a sample of data is needed in order to attain a desired level of accuracy in our estimates.

SUPPLEMENTARY EXERCISES

1 Proposed SEC regulations state that an auditor using a random sampling must be 99% sure of having one class of companies' mean amount of accounts receivable estimated correctly to within 5 dollars. If experience has shown that accounts receivable of this type of company generally have standard deviation of 60 dollars, how many accounts does the auditor have to include in his random sample in order to achieve the specified accuracy?

2 In connection with a study on voting patterns, a political scientist would like to choose a random sample of voters in order to estimate the mean yearly income level for a certain precinct. It is known from earlier studies of this kind that such incomes have standard deviation 3000 dollars in a given precinct. If the political scientist wants to be 90% sure of having the true mean income estimated correctly to within 100 dollars, how many voters does he need for the random sample?

3   The research branch of a major automobile manufacturer has been working on an engine that is supposed to be able to get more miles per gallon of gasoline. In 9 trial runs during a preliminary testing period, the engine was used on a 100-mile track, and the amount of gasoline (in gallons) needed to cover the route was recorded each time as follows:

2.5   3.0   2.0   2.4   2.9   2.6   2.7   2.1   2.3

(a) Find a 90% confidence interval for the engine's true mean consumption of gasoline per 100 miles.
(b) If it is reasonable to use .35 as an estimate of the standard deviation, how many trial runs would be needed to be able to estimate the true mean consumption to within .1 gallon per 100 miles with 98% confidence?

4   A political science researcher studying the loss of seats in the U.S. House of Representatives to the party in power at off-year elections found that in 36 such elections the party in power lost an average of 11.5 seats with a standard deviation of 5. With what confidence can we assert that the true mean loss to the party in power is somewhere between 9 and 14 seats?

5   A sociological survey involving 400 randomly selected families in one section of a large city showed that the average family surveyed had 2.37 children with a standard deviation of 1.80 children. Find a 95% confidence interval for the true mean number of children per family in that section of the city.

6   As part of a study of highway safety, a random sample of 25 automobiles from the freeways of Los Angeles were tested for their stopping ability after application of the brakes. The 25 cars required a mean distance of 148 feet to stop with a standard deviation of 21 feet. Find a 90% confidence interval for the true mean stopping distance of all cars on Los Angeles freeways.

7   In connection with a biological study of the growth of cherry tomato plants, 60 plants had produced an average of 14.8 pounds of tomatoes with a standard deviation of 6.5 pounds after 100 days of growth.
(a) Find a 90% confidence interval for the true mean cherry tomato yield, in pounds.
(b) Find a 98% confidence interval for the mean yield.

8   As part of an analysis of the relationship between smoking and lung disease, a public health physician conducts a survey of hospital patients with lung disorders, with the goal of finding out what proportion of such patients are heavy smokers.
(a) Of 200 such patients, the physician determines that 120 are heavy smokers. Find a 95% confidence interval for the true proportion of lung patients who are heavy smokers.
(b) If an expanded survey reveals that 1200 out of 2000 patients are heavy smokers, what would be a 95% confidence interval for the true proportion?
(c) If the physician wants to be 95% sure of having the true proportion estimated correctly to within .01 (1%), how many patients must he include in his survey?

9 Of 300 licensed drivers selected at random from the Department of Motor Vehicles records, it is learned that 45 are problem drinkers. Construct a 90% confidence interval for the true proportion of licensed drivers who are problem drinkers.

10 In an election in which Able is running against Baker, a survey of 1500 voters shows that 780 intend to vote for Able.
   (a) Find an 80% confidence interval for the true proportion of the vote going to Able.
   (b) Find a 90% confidence interval for the true proportion of the vote going to Able.

11 To estimate the proportion of a company's accounts that have defaulted over the years, an accountant chooses a random sample of 2000 accounts and discovers that 60 have defaulted.
   (a) Find a 95% confidence interval for the true proportion of defaulting accounts.
   (b) If it were necessary to be 99.9% sure of having the true proportion estimated to within .01, how large a random sample would be needed to guarantee that level of accuracy?

4

TESTING
STATISTICAL
HYPOTHESES

One of the fundamental principles of the scientific method of increasing our knowledge of the world around us is the formulation of a hypothesis.[1] Today this and other principles of the scientific method have just as much validity for the social and managerial sciences as they do for the natural sciences. When we formulate a hypothesis, we define very clearly the problem under discussion and we provide the framework for gathering the relevant facts. An analysis of the accumulated facts (or data) then leads us to accept or reject the hypothesis.

For example, an educational psychologist might want to know whether pupils who have participated in the Head Start program at an early age do better in sixth-grade mathematics than those from similar backgrounds who have not participated. One way she could set up the problem would be to define the hypothesis:

[1]Some individuals use the terminology, "an hypothesis," by analogy with "an honor," "an historian," "an horse."

H: Head Start participants do *no better* in sixth-grade mathematics than those from similar backgrounds who have not participated

and the alternative

A: Head Start participants do *better* than those who have not participated

The next step would be to choose a random sample of sixth-grade pupils who have participated and a random sample of those who have not, and to give all pupils selected an examination in sixth-grade mathematics. A comparison of the examination scores of the two groups would then lead the psychologist to reject or accept the hypothesis **H**. By rejecting the hypothesis **H** in favor of the alternative **A**, she would be concluding that the Head Start program tends to improve mathematics ability at the sixth-grade level; by failing to reject **H**, she would be concluding that it does not.

Our objective in this chapter is to show how to formulate a hypothesis in an applied situation and how to use statistical methods of data analysis for the purpose of deciding whether to reject or accept that hypothesis.

SECTION 4.A
THE STATISTICAL DECISION-MAKING PROCESS

Suppose an individual complaining of stomach pains comes to a doctor's office. Mentally, the doctor immediately formulates the hypothesis

H: The person has appendicitis

and the alternative

A: The person has a stomachache

The doctor's job is to decide between these two possibilities. We can run some medical tests in this and any other situation, for example, blood tests, urine samples, x-rays, etc. After all the data are assembled, the individual in question fits into one of the following categories:

1 He almost certainly has appendicitis.
2 He probably has appendicitis.
3 The evidence is inconclusive.
4 He probably does not have appendicitis.
5 He almost certainly does not have appendicitis.

Figure 4.1 illustrates the situation. There is no such thing as absolute certainty in

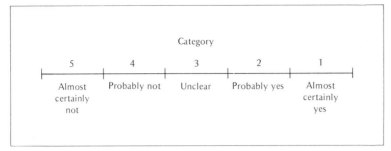

FIGURE 4.1
Categories of individual in regard to having appendicitis.

a situation like this until it is almost too late—emergency appendectomies are not very unusual operations.

TWO TYPES OF ERRORS

After we arrive at our decision concerning the medical state of the individual in question, we may have made one of two types of errors. These types of errors are formally designated "Type I" and "Type II" and they are as follows:

Type I error: We reject **H** when it is really true. (In this case, we decide that the individual has a stomachache when he in fact has appendicitis.)

Type II error: We accept **H** when it is really false. (In this case, we decide that the individual has appendicitis when he in fact has a stomachache.)

Of course, we might possibly arrive at the correct decision: We could reject **H** when it's really false or accept it when it's really true.

We have to settle two questions in connection with Type I and Type II errors: (1) Which of the two errors is worse? (2) How often are we willing to make the worse of the two errors?

Regarding the appendicitis problem, we would probably agree that the Type I error is worse. If we make a Type I error, we are risking the life of the patient, while if we make a Type II error, we are merely wasting his money, the insurance company's money, and the medical staff's time.

How often would we be willing to make a Type I error? Well, I don't know about you, but if I were the doctor, I'd prefer never to make a Type I error. What decision strategy can we follow that would guarantee our never making a Type I error? If you think about it, you'll see that the only way we can be 100% sure of never making a Type I error would be to always decide that the person has appendicitis and then to operate accordingly. This way, we can be sure of never rejecting **H** when it's really true because we will never reject **H**—never when it's true and never when it's false.

Imagine the situation for a moment. The individual comes to our office saying, "My tummy hurts," and before he finishes the sentence, we're rushing him off to the appendectomy ward. Although we are guaranteed never to make a Type I error, you'll have to agree that this diagnostic technique is somewhat impractical. Why is this procedure impractical? Primarily because we will be doing appendectomies on many, many people who do not have appendicitis. More precisely, in our zeal to avoid Type I errors, we will be committing Type II errors whenever it is possible to do so. If we consider the fact that probably fewer than one-tenth of 1% of persons with stomach pains actually have appendicitis, it's ridiculous to perform an appendectomy on everyone who walks through the door.

Well, now we're back where we started. We agree that we cannot operate on everyone, and therefore we're going to have to prescribe Pepto-Bismol for some of our patients. Yet, every time we let a person slip through with a bottle of Pepto-Bismol instead of an appendectomy, we're risking a Type I error. Even if we let only one person through without an appendectomy, there is some chance that the person in question actually has appendicitis. What do we do? How many people do we let by? How many do we operate on?

The problem revolves around the difficulty of deciding upon an acceptable level of Type I errors. How often are we willing to make a Type I error; i.e., how many persons who really have (unknown to us at the time) appendicitis are we willing to send home with a bottle of Pepto-Bismol?

Of course, there is no real answer to this question. This is a subjective decision that must take many factors into account. If we denote by α (pronounced "alpha") our probability of making a Type I error and by β (pronounced "baytah") our probability of making a Type II error, there is an inverse relationship between α and β. If α is very low, we would not be making a Type I error too often. Therefore we would not be rejecting **H** (and consequently asserting **A** to be true) unless it were fairly certain that **H** was false. It would therefore be necessary to accept **H** in some questionable cases, and so our probability of a Type II error would be high. So a low value of α would almost automatically lead to a high value of β. The reverse is also true: A high value of α means we are rejecting **H** in questionable cases, so that we accept **H** only when it is probably true and β is therefore low. It is therefore necessary for us to weigh the relative seriousness of the two errors, not merely to decide separately on an acceptable level of Type I errors. As we have seen in the appendicitis example above, the insistence on a low incidence of Type I errors almost certainly leads to an unacceptably high level of Type II errors. On the other hand, we can't really accept too high a level of Type I errors either.

To make the best of a bad situation, we use the following procedure for handling statistical hypotheses:

1 We decide which of the two errors, Type I or Type II, is worse.
2 If we think the Type I error is worse, we choose a testing procedure which

makes α acceptably low, and we agree to be satisfied with whatever the value of β turns out to be.

3 If we think the Type II error is worse, we choose a testing procedure which makes α high, and then we can expect β to be correspondingly low.

Despite the general rule of the inverse behavior of α and β, it unfortunately turns out that there is usually not a simple mathematical relationship between α and β. The only thing we can really be sure of is that as α increases from low to high, β will decrease from high to low. For this reason, it is advisable whenever possible to construct **H** and **A** in such a way that the Type I error will be worse, and then our ability to choose α will give us some control over that error.

THE LEVEL OF SIGNIFICANCE

Our choice of α can therefore be viewed as a subjective measure, on our part, of the relative seriousness of the two types of errors. If we think the Type I error is much more serious in a particular applied context, we should choose α to be low. If we think the Type II error is very serious, we should choose α to be high. For this reason, α is said to be the "significance level of the test."

Any decision that is made on the basis of a statistical analysis must include a statement of the numerical value of α, for only in this way can a reader of the report know how the writer viewed the relative seriousness of the two errors. In some borderline cases, a change in the numerical value of α may very well result in a change in the decision.

You have probably heard or read somewhere the allegation that anything can be proved using statistics. The greater part of whatever truth there may be to this allegation arises from disagreements between opposing parties in regard to the appropriate level of significance to be used in analyzing the data. As a prime example of this sort of situation, we can look at the recent history of disputes between environmentalists and economic growth advocates over offshore oil drilling in locations such as California's Santa Barbara Channel. The hypothesis to be tested is

H: Offshore oil drilling *does not* cause significant environmental damage

and the alternative is

A: Offshore oil drilling *does* cause significant environmental damage

In this situation, the significance level would be

$\alpha = P(\text{Type I error}) = P(\text{rejecting } \mathbf{H} \text{ when it's really true})$

= probability of asserting that offshore oil drilling *does* cause significant environmental damage *when*, in fact, it *does not* while

$$\beta = P(\text{Type II error}) = P(\text{accepting } \mathbf{H} \text{ when it's really false})$$

$$= \text{probability of asserting that offshore oil drilling } does \; not \text{ cause significant environmental damage } when, \text{ in fact, it } does$$

From the environmentalists' point of view, the Type II error would be worse, for the environmentalist believes that the possibility of significant damage to the environment outweighs any economic advantage to be gained from offshore oil drilling. In the opinion of advocates of economic growth, the Type I error is the worse of the two because such an error would tend to stagnate economic growth, leading to loss of jobs and a lowered standard of living, even if there were no real evidence that significant environmental damage would occur. It is not that environmentalists are necessarily out to wreck the economy, nor are economic growth advocates deliberately out to destroy the environment. What is really the case is that, if an error (Type I or Type II) is to be made, the environmentalist would rather err on the side of the environment (while willing to risk some economic dislocation), and the economic growth advocate would rather err on the side of economic well-being (while willing to risk some environmental problems).

Environmentalists would then base a statistical analysis on a high level of significance α, because they would insist that $\beta = P(\text{Type II error})$ be held as low as possible. Growth advocates, on the other hand, would favor a low level of significance, for they would feel that $\alpha = P(\text{Type I error})$ itself should be low. We summarize this discussion in Table 4.1.

One more comment before leaving this section. What do we mean by a "low" α, and what do we mean by a "high" α? Because the choice of α is a subjective decision of the researcher or manager on the scene, this question cannot be answered using a mathematical formula. An examination of professional publications in various applied fields indicates that a high level of α usually means a value between .02 and .10, while a low level of α is taken to be between .005 and .02. If the Type I error is worse, the typical value of α is .01, while if the Type II error is worse, α is typically taken to be .05.

TABLE 4.1
Choice of level of significance α—environmentalists versus economic growth advocates

H: Drilling does not cause environmental damage
A: Drilling does cause environmental damage

		Significance Level	
Primary Concern	Worse Error	α	β
Environment	Type II	High	Low
Economy	Type I	Low	High

EXERCISES 4.A

1 Many statistical studies have been conducted in an attempt to find out whether or not smoking really causes cancer. To put the burden of proof on those who feel that it does, we can formulate the hypothesis

    H: Smoking does not cause cancer

and the alternative

    A: Smoking causes cancer

  (a) The cigarette industry would be unfairly hurt by a definitive decision that smoking causes cancer if the opposite were, in fact, true. To protect the cigarette industry, should α be chosen high or low?

  (b) The American Cancer Society wants to eliminate all possible causes of cancer even if their connection with cancer is not conclusively proven beyond all doubt. Should the Society argue for a high or low value for α?

2 A jury is suppose to convict an individual on trial for a string of murders and assaults if the individual is guilty beyond a reasonable doubt. The word "reasonable" is related to the numerical value of α in testing the hypothesis

    H: The individual is innocent

and the alternative

    A: The individual is guilty

  (a) To avoid convicting an innocent person, should the jury choose a high value for α or a low value?

  (b) To avoid releasing a guilty person to possibly continue the wave of attacks, should the jury select a high value for α or a low value?

3 A business executive has been assigned the task of deciding whether or not to invest several thousand dollars in a new business venture. He has to choose between the hypothesis

    H: The venture will succeed

and the alternative

    A: The venture will fail

  (a) If there is a shortage of investment capital, so that only the safest investments should be undertaken, should the executive use a high value or a low value of α?

(b) If investment capital is plentiful, allowing the funding of some possibly risky ventures, should α be chosen as high or low?

SECTION 4.B
ONE-SAMPLE TESTS FOR THE MEAN

Because of the expense involved in manufacturing Nitro-Plus, a new chemical fertilizer, the fertilizer must produce an average yield of more than 20,000 pounds of tomatoes per acre within a specified growing period in order to be economically feasible. An independent agricultural testing organization chose 40 acres of farmland at random in various geographical regions, fertilized each with Nitro-Plus, and planted tomatoes intensively. The results show that the 40 acres had a mean yield of 20,400 pounds, and the yield per acre varied widely, having a standard deviation of 1200 pounds. On the basis of the test results involving 40 randomly selected acres, can we conclude that the *true* mean yield really exceeds 20,000 pounds per acre?

In order to put the burden of proof on the fertilizer, we denote the true mean yield by the symbol μ, and we define the hypothesis to be

H: $\mu = 20,000$ (the true mean yield fails to exceed 20,000)

and the alternative

A: $\mu > 20,000$ (the true mean yield exceeds 20,000)

What is a reasonable significance level for this test? Well, in this case, the Type I error (rejecting **H** when it's really true) would be to conclude that the fertilizer produced the desired level of yield when, in fact, it did not. The Type II error, on the other hand, would be to say that Nitro-Plus does not meet the specifications when, in fact, it really does. We might be able to agree that the Type I is the worse error from the farmer's viewpoint, for it would eventually lead to financial failure of the operation, while the Type II would only prevent us from taking advantage of an effective new fertilizer. Suppose, therefore, we set the significance level low, say $\alpha = .01$.

LARGE SAMPLES

Now, how do we use the data to decide whether to accept or reject the hypothesis **H**? If we recall the statement of the central limit theorem that appears in Sec. 3.B, we know that the collection of all possible sample means $\bar{x}$ of size $n = 40$ constitutes a set of approximately normally distributed data having mean μ and standard deviation $s/\sqrt{n}$. More precisely, we know that for all possible samples composed of $n = 40$ data points, the numbers

$$z = \frac{\bar{x} - \mu}{s/\sqrt{n}}$$

have approximately the standard normal distribution. Assuming that $\mu = 20{,}000$ (namely, that **H** is *really* true), under what circumstances would we be misled into rejecting **H**? We would tend to reject **H** if the sample mean $\bar{x}$ came out to be substantially larger than $\mu = 20{,}000$.

What do we mean by "substantially larger"? Well, this is the point at which the significance level of the test enters the picture. As we have noted that a reasonable level of significance here would be $\alpha = .01$, this means that we want to reject **H** only 1% of the time when it, **H**, is really true. Because we would tend to reject **H** if $\bar{x}$ turned out to be much larger than $\mu = 20{,}000$, the rejection of **H** would be indicated by a large value of

$$z = \frac{\bar{x} - 20{,}000}{s/\sqrt{n}}$$

because z would be large when $\bar{x}$ is much larger than $20{,}000$. (z is actually a measure of the difference between $\bar{x}$ and $20{,}000$.) Now, when **H** is really true, there is a 1% chance that z would exceed $z_{.01} = 2.33$. Therefore, if we agree to reject **H** in case z turns out to be larger than 2.33, then we will have only a 1% chance of rejecting the hypothesis **H** that $\mu = 20{,}000$ when it is really true. This means that our test of whether or not the true mean yield per acre of tomatoes using Nitro-Plus fertilizer really exceeds 20,000 pounds will have significance level $\alpha = .01$.

The set of values of z which lead us to reject **H** is called the "rejection region" of the test. Therefore, the rejection region of this test is the region where $z > 2.33$. A diagram with the rejection region shaded appears in Fig. 4.2. We can

FIGURE 4.2
Rejection region for H when $\alpha = .01$ (tomatoes-fertilizer example).

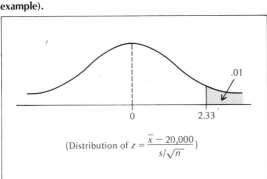

formalize the hypothesis testing mechanism for this problem as follows:

H: $\mu = 20,000$

A: $\mu > 20,000$

Compute $z = (\bar{x} - 20,000)/(s/\sqrt{n})$ and reject **H** at level $\alpha = .01$ if $z > z_{.01} = 2.33$. To complete this problem and to make our decision regarding the economic feasibility of the Nitro-Plus fertilizer, it remains only to calculate z and to decide whether or not it exceeds 2.33.

From the data accumulated by the independent testing organization, a study of $n = 40$ randomly selected acres gave a sample mean yield of $\bar{x} = 20,400$ with a sample standard deviation of $s = 1200$. Inserting these numbers into the formula for z, we find that

$$z = \frac{\bar{x} - 20,000}{s/\sqrt{n}} = \frac{20,400 - 20,000}{1200/\sqrt{40}} = \frac{400}{1200/6.325} = \frac{(400)(6.325)}{1200}$$

$$= 2.11$$

We agreed to reject **H** if $z > 2.33$, but z actually turned out to be 2.11 (which is less than 2.33). We therefore *cannot* reject **H**. Therefore, at significance level $\alpha = .01$, we cannot conclude that the true mean yield exceeds 20,000 pounds per acre. Our experimental yields have not been large enough to provide convincing proof of the economic feasibility of Nitro-Plus.

Suppose we had looked at the problem from a different point of view. Suppose the economic goal of running a profitable operation had been outweighed by the necessity of producing a larger food supply as soon as possible. In such a situation, we would probably be less concerned with the precise level of mean yield and more concerned merely with the ability of Nitro-Plus to produce a large yield. In testing **H** against **A**, we would therefore be likely to consider the Type II error (accepting **H** when it's really false) to be the worse for it would lead us into rejecting the use of Nitro-Plus in some cases when the fertilizer would be truly effective. In our desire to avoid a Type II error, we would then be willing to ease up somewhat on the probability of a Type I error, perhaps by raising the significance level from $\alpha = .01$ to $\alpha = .05$. At significance level $\alpha = .05$, our rejection rule would be to reject **H** in favor of **A** if

$$z > z_{.05} = 1.64$$

The rejection region is illustrated in the diagram of Fig. 4.3.

In viewing the problem from this different point of view, the only thing that has changed is the level of significance α. The data, of course, remain the same. (Our opinion of the problem will have no effect on the data, for the data record only how many pounds of tomatoes are actually produced using the Nitro-Plus

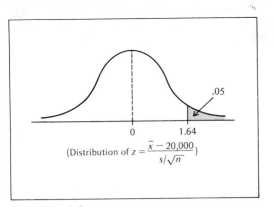

(Distribution of $z = \frac{\bar{x} - 20{,}000}{s/\sqrt{n}}$)

FIGURE 4.3
Rejection region for H when $\alpha = .05$ (tomatoes-fertilizer example).

fertilizer.) Therefore, we still have $n = 40$, $\bar{x} = 20{,}400$, and $s = 1200$, so that

$$z = \frac{\bar{x} - 20{,}000}{s/\sqrt{n}} = \frac{20{,}400 - 20{,}000}{1200/\sqrt{40}} = 2.11$$

as before. Now, however, we have agreed to reject **H** if $z > 1.64$. Because $z = 2.11$ has turned out to be larger than 1.64, we would therefore make the decision to reject the hypothesis **H** in favor of alternative **A**.

Our conclusion would then be that Nitro-Plus increased tomato output to an extent sufficient to justify its use. At significance level $\alpha = .05$, our experimental results indicate that Nitro-Plus will be able to produce yields in excess of 20,000 pounds per acre. A tabular comparison between the cases $\alpha = .01$ and $\alpha = .05$ is presented in Table 4.2.

The statistical test we have just carried out in our analysis of the tomatoes-fertilizer example is called the "one-sample z test." It is a one-sample test because only one set of data is used in the calculations, and it is a z test because the availability of more than 30 data points (here $n = 40$) allows us to apply the central limit theorem for the purpose of using percentage points of the standard normal distribution. One-sample tests for the mean of a population are characterized by the use of three distinct means: the true mean μ of the population, the

TABLE 4.2
Decision making in the tomatoes-fertilizer example
H: $\mu = 20{,}000$ versus **A:** $\mu > 20{,}000$

Significance Level	Rejection Rule	Calculated Value of z	Decision
$\alpha = .01$	$z > 2.33$	$z = 2.11$	Accept **H**
$\alpha = .05$	$z > 1.64$	$z = 2.11$	Reject **H**

sample mean $\bar{x}$ based on the data, and finally the "hypothesized mean" which we shall denote by the symbol μ_0 ("mew-zero"). In the tomatoes-fertilizer example, the hypothesis reads

H: $\mu = 20{,}000$

This statement does not really say that the true mean μ is 20,000, but merely that we are hypothesizing that it is. We therefore refer to 20,000 as the hypothesized mean, and we denote it by μ_0. The sample mean is $\bar{x} = 20{,}400$, and the numerical value of the true mean μ is, as always, unknown to us.

The format of the one-sample z test is often as it was in the tomatoes-fertilizer example. In particular, we test the hypothesis

H: $\mu = \mu_0$

against the alternative

A: $\mu > \mu_0$

we compute

★ $$z = \frac{\bar{x} - \mu_0}{s / \sqrt{n}}$$ (4.1)

and we reject **H** at level α if $z > z_\alpha$. Sometimes, however, the alternative is formulated differently due to the nature of the question being asked. The following example provides an illustration.

Example 4.1 Calculator Battery Lifetimes One manufacturer of pocket calculators advertises that its battery pack allows the calculator to operate continuously for 22 hours on the average without recharging. A prospective corporate customer tested a sample of 50 calculators, and these turned out to have mean 21.8 hours with a standard deviation of .9 hour. If a significance level of $\alpha = .10$ seems to be appropriate, is there sufficient evidence to indicate that the true mean duration of continuous operation (without recharging) is actually less than the 22 hours claimed by the manufactuer?

SOLUTION Denoting by μ the true mean duration of continuous operation (without recharging) of the battery pack, we have to test the hypothesis

H: $\mu = 22$

versus

A: $\mu < 22$

at the level of significance $\alpha = .10$. Because we have a single sample of data and $n = 50 > 30$ data points, the appropriate method of analysis is again the one-sample z test. We compute, as before,

$$z = \frac{\bar{x} - 22}{s/\sqrt{n}}$$

and we have to determine which values of z will lead us to reject **H** in favor of **A**. We will be inclined to reject **H** if $\bar{x}$ turns out to be considerably smaller than 22, relatively speaking, because if our sample mean comes out quite a bit lower than 22, we'd have a hard time convincing ourselves that the true mean was really 22. Yet, as $\alpha = .10$ here, we would be making a Type I error the 10% of the time z actually falls below $-z_{.10} = -1.28$ even when the true mean is really 22. Therefore, if we decide to reject **H** in favor of **A** in case z turns out to be less than (even more negative than) $-z_{.10} = -1.28$, our test would have significance level $\alpha = .10$, indicating a 10% chance of a Type I error. We therefore agree to reject **H** at level $\alpha = .10$ in this situation if $z < -1.28$. A diagram of the rejection region is presented in Fig. 4.4.

It now remains only to calculate the numerical value of z and to determine whether or not it is more negative than -1.28. From the data, we see that $n = 50$, $\bar{x} = 21.8$, and $s = .9$.

We therefore calculate z as follows:

$$z = \frac{\bar{x} - 22}{s/\sqrt{n}} = \frac{21.8 - 22}{.9/\sqrt{50}} = \frac{-.2}{.9/7.07} = \frac{(-.2)(7.07)}{.9} = -1.57$$

As $z = -1.57 < -1.28$, we have to reject **H** in favor of **A**. At level $\alpha = .10$, our conclusion is that the mean duration of continuous operation (without recharg-

FIGURE 4.4
Rejection region for H when $\alpha = .10$ (calculator battery example).

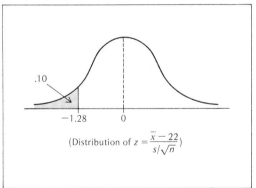

ing) of the battery pack is actually *less than* the 22 hours claimed by the manufacturer.

SMALL SAMPLES

The two situations studied here, namely, the tomatoes-fertilizer example and the
calculator battery example, called for the use of the one-sample z test. The z in
the one-sample z test refers to the fact that more than 30 data points are available, so that we can base our analysis of the problem on the z scores of the
normal distribution. What happens in a one-sample problem where fewer than
30 data points are avilable? A brief glance at the second half of Sec. 3.B should
provide the answer. In situations where the underlying population (from which
the sample of data points is randomly selected) is normally distributed,[1] we can
replace the rejection regions based on the z scores of Table A.3 of the Appendix
by rejection regions based on the t distribution of Table A.4.

Example 4.2 Body Temperatures A medical doctor who is also an amateur
anthropologist is interested in finding out whether the average body temperature of Alaskan Eskimos is significantly lower than the usual American average,
which is 98.6°F. Eight Eskimos selected at random from state of Alaska census
lists had the following recorded body temperatures in degrees:

$$98.5 \quad 98.1 \quad 98.6 \quad 98.7 \quad 98.4 \quad 98.9 \quad 98.0 \quad 98.4$$

At significance level $\alpha = .05$, do the results of the study support the assertion
that Eskimos really have lower body temperature than Americans who are natives of warmer climates?

SOLUTION   Because $n = 8 < 30$, the solution of this problem requires the use of
the "one-sample t test". If we denote by μ the true mean body temperature of
Alaskan Eskimos, we want to test the hypothesis

> **H:** $\mu = 98.6$ (Eskimos have the same average body temperature as other
> Americans)

against the alternative

> **A:** $\mu < 98.6$ (Eskimos have lower body temperature)

at the level of significance $\alpha = .05$. If we were using the z test, we would reject
H if $z < -z_{.05} = -1.64$. However, as we are using the t test, we replace $z_{.05}$ by
$t_{.05}[n- 1]$, where $n - 1$ is the degrees of freedom required in using Table A.4.

---

[1]If it's not, see Chap. 8.

TABLE 4.3
A small portion of Table A.4

df	$t_{.05}$
7	1.895

Therefore, as $n = 8$, we have df $= n - 1 = 8 - 1 = 7$. We compute

★ $$t = \frac{\bar{x} - \mu_0}{s/\sqrt{n}} \qquad\qquad (4.2)$$

in which $\mu_o = 98.6$ here, and we reject **H** in favor of **A** if $t < -t_{.05}[\,7\,] = -1.895$. We illustrate the relevant portion of Table A.4 in Table 4.3 and the rejection region in Fig. 4.5.

It remains, then, only to compute the value of

$$t = \frac{\bar{x} - 98.6}{s/\sqrt{n}}$$

and to check whether or not $t < -1.895$. We can insert the value $n = 8$ immediately, but the numerical values of $\bar{x}$ and s will have to be computed directly from the data. As in Chaps. 1 and 3, we will set up a table of calculations to do the job. The calculations, which can be found in Table 4.4, show that $\bar{x} = 98.45$ and $s = .3$. Substitution of these numbers into the formula for t yields

$$t = \frac{\bar{x} - 98.6}{s/\sqrt{n}} = \frac{98.45 - 98.6}{.3/\sqrt{8}} = \frac{-.15}{.3/2.83} = \frac{(-.15)(2.83)}{.3} = -1.42$$

FIGURE 4.5
Rejection region for *H* when $\alpha = .05$ (body temperature example).

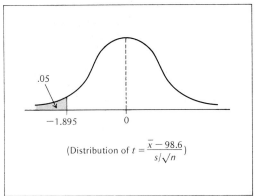

(Distribution of $t = \dfrac{\bar{x} - 98.6}{s/\sqrt{n}}$)

TABLE 4.4
Calculations leading to the mean and standard deviation of the body temperature data

x	$x - \bar{x}$	$(x - \bar{x})^2$
98.5	.05	.0025
98.1	−.35	.1225
98.6	.15	.0225
98.7	.25	.0625
98.4	−.05	.0025
98.9	.45	.2025
98.0	−.45	.2025
98.4	−.05	.0025
Sums 787.6	.00	.6200

$n = 8$

$$\bar{x} = \frac{\Sigma x}{n} = \frac{787.6}{8} = 98.45$$

$$s = \sqrt{\frac{\Sigma(x - \bar{x})^2}{n - 1}} = \sqrt{\frac{.62}{7}} = \sqrt{.0886} = .3$$

Now, recall that we agreed to reject **H** in favor of **A** if t turned out to be less than −1.895. Well, as it turns out, $t = −1.42$ is *not* less than −1.895 because it is less negative and, therefore, larger. Therefore, since $t = −1.42 > −1.895$, we *cannot* reject **H** in favor of **A**. We conclude at level $\alpha = .05$ that the data *do not support* the assertion that Eskimos have lower body temperatures, on the average, than Americans of warmer climates.

Example 4.3 Depth Perception A psychologist, conducting a study of the average person's ability to judge distances, sets up a test of depth perception in which randomly selected individuals attempt to estimate the distance between two markers. The markers were actually 2.5 feet apart, and the 10 participants in the study gave the following estimates:

2.1　　1.8　　2.3　　2.3　　2.6　　2.5　　2.3　　2.5　　2.1　　2.5

At level $\alpha = .05$, do the results of the study indicate that persons have difficulty in accurately estimating the correct distance?

SOLUTION　A one-sample t test is required for the solution of this problem, in view of the fact that we have fewer than 30 data points available, namely, 10. We use μ to represent the true mean estimate of the distance between the markers by the entire population in question. We then test the hypothesis

H:　$\mu = 2.5$ (on the average, persons can accurately estimate the distance between the markers)

versus

A: $\mu \neq 2.5$ (on the average, persons have difficulty in accurately estimating the distance between the markers)

Under what circumstances would we feel compelled to reject **H** in favor of **A**? Well, as $\bar{x}$ is an approximation of μ, we would be inclined to reject **H** if the number

$$t = \frac{\bar{x} - 2.5}{s/\sqrt{n}}$$

turned out to be either too large or too small because both of these extremes would tend to indicate that μ is relatively far from 2.5.

For a significance level of $\alpha = .05$, we are willing to make a Type I error 5% of the time; this means that a rejection region of the sort pictured in Figs. 4.2, 4.3, 4.4, and 4.5 must have a shaded region representing a probability of .05. In the present example, however, the rejection region would have to appear in two parts, one part at each extreme of the graph of the t distribution. It would make sense to organize the rejection region in such a way that the far right part and the far left part each account for 2.5% of the possible values of t. The alternative **A:** $\mu \neq 2.5$ does not allow us to favor one extreme over the other. We would therefore be using the 2.5 percentage point of the t distribution, the number $t_{.025}[n-1] = t_{.025}[9] = 2.262$, to yield our points of division between the rejection and acceptance regions. A small portion of Table A.4 is reproduced in Table 4.5 to point up the selection of $t_{.025}[9]$. We would therefore reject **H** at level $\alpha = .05$ if either t were more negative than $-t_{.025}[9] = -2.252$ or if t were greater than $t_{.025}[9] = 2.262$. In other words, we would compute

$$t = \frac{\bar{x} - 2.5}{s/\sqrt{n}}$$

and we would reject **H** in favor of **A** if $t < -2.262$ or if $t > 2.262$. We can express the two-part rejection rule more compactly in terms of the absolute value of t: We reject **H** if $|t| > 2.262$. The rejection region is illustrated in the diagram of Fig. 4.6.

The preliminary calculations appearing in Table 4.6 are aimed at obtaining the numerical values of $\bar{x}$ and s directly from the psychologist's $n = 10$ data

TABLE 4.5
A small portion of Table A.4

df	$t_{.05}$	$t_{.025}$
9	1.833	2.262

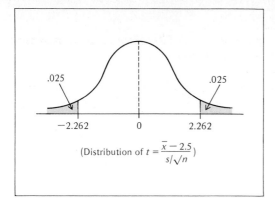

$$(\text{Distribution of } t = \frac{\bar{x} - 2.5}{s/\sqrt{n}})$$

FIGURE 4.6
Rejection region for *H* when $\alpha = .05$ (depth perception example).

points. From those calculations, we discover that $\bar{x} = 2.3$ and $s = .245$. Inserting these numbers into the formula for t, we get

$$t = \frac{\bar{x} - 2.5}{s/\sqrt{n}} = \frac{2.3 - 2.5}{.245/\sqrt{10}} = \frac{-.2}{.245/3.162} = \frac{(-.2)(3.162)}{.245} = -2.58$$

We agreed to reject **H** if $|t| > 2.262$. Because $t = -2.58$, it follows that $|t| = 2.58$ which indeed exceeds 2.262, and we should therefore reject **H**. We conclude at level $\alpha = .05$ that persons generally have difficulty in accurately measuring distance, because our analysis of the psychologist's study reveals that, for markers placed 2.5 feet apart, the average person will tend to estimate the distance as being significantly different from 2.5.

The depth perception example has provided an illustration of what is called a "two-tailed t test" (not to be confused with the "two-taled tee test" used in resolving disputes over golfing scores). A two-tailed test is a statistical test having a two-part rejection region of the sort pictured in Fig. 4.6. Such a rejection region arises in tests of population means when the alternative is of the form

A: $\mu \neq \mu_0$

rather than **A:** $\mu > \mu_0$ or **A:** $\mu < \mu_0$. For a two-tailed test having significance level α, we reject **H** if $|t| > t_{\alpha/2} [n - 1]$, using $\alpha/2$ rather than α to locate the rejection region. For quick reference in dealing with situations like those discussed in the present section and also those of the next two sections, we can collect the various possible alternatives together with the rejection rules corresponding to them. This listing of possible alternatives appears in Table 4.7, and use of that table will make it unnecessary to draw diagrams like those of Figs.

TABLE 4.6
Calculations leading to the mean and standard deviation of the depth perception data (shortcut formula)

x	x^2
2.1	4.41
1.8	3.24
2.3	5.29
2.3	5.29
2.6	6.76
2.5	6.25
2.3	5.29
2.5	6.25
2.1	4.41
2.5	6.25
23.0	53.44

$n = 10$

$$\bar{x} = \frac{\Sigma x}{n} = \frac{23.0}{10} = 2.3$$

$$s = \sqrt{\frac{n\Sigma x^2 - (\Sigma x)^2}{n(n-1)}} = \sqrt{\frac{10(53.44) - (23.0)^2}{(10)(9)}}$$

$$= \sqrt{\frac{534.4 - 529}{90}} = \sqrt{\frac{5.4}{90}} = \sqrt{.06} = .245$$

4.2, 4.3, 4.4, 4.5, and 4.6 for each and every problem that comes up. Table 4.7 incorporates the ideas that went into the determination of each of the rejection regions pictured above. We shall have occasion to refer to Table 4.7 during the next two sections of the present chapter and also during parts of Chap. 8.

TABLE 4.7
Rejection rules in the presence of various alternatives for tests involving population means (**H**: $\mu = \mu_0$)

Alternative	Reject **H** in Favor of **A** at Significance Level α if:
A: $\mu > \mu_0$	$z > z_\alpha$
A: $\mu < \mu_0$	$z < -z_\alpha$
A: $\mu \neq \mu_0$	$|z| > z_{\alpha/2}$

NOTE: Although the above information is given in the language of the one-sample z test, it is necessary merely to replace the z_α and $z_{\alpha/2}$ by $t_\alpha[n-1]$ and $t_{\alpha/2}[n-1]$ in order to obtain the rejection rules for the one-sample t test.

EXERCISES 4.B

1 In the 1950s, the average number of children per family in the United States was 2.35. One sociologist is interested in finding out whether the number of children per family has declined from that level during the present decade. He selects a random sample of 81 families from around the nation and discovers that they have 2.17 children on the average with a standard deviation of .9 child. Do the data support at level $\alpha = .05$ the contention that families of the 1970s have on the average fewer children than families of the 1950s?

2 An automobile manufacturer claims that his cars use an average of 4.50 gallons of gasoline for each 100 miles. A consumer organization tests 36 of the cars and finds that a mean of 4.65 gallons with a standard deviation of .36 gallon is used for each 100 miles by the cars tested. At significance level $\alpha = .01$, do the data provide evidence that the cars really use more than 4.50 gallons on the average per 100 miles?

3 A machine which puts out plastic piping is, according to contractual specifications, supposed to be producing piping of diameter 4 inches. Piping which is either too large or too small will not meet the construction requirements. As part of a routine quality control test, a random sample of 50 pieces were carefully measured, and their diameters had mean 4.05 inches and standard deviation .12 inch. At level $\alpha = .05$, can we conclude that the true mean diameter differs significantly from 4 inches?

4 It is the policy of one school district to hire credentialed reading specialists whenever the true mean reading score of the district's sixth-graders falls below 40, as measured by a particular standard test. If a random sample of 25 pupils had mean reading score 38 with a standard deviation of 3, is the hiring of reading specialists justified
   (a) At significance level $\alpha = .05$?
   (b) At significance level $\alpha = .005$?

5 A tobacco processor certified that his cigarettes contained no more than 21 milligrams of nicotine on the average. A random sample of 25 cigarettes was analyzed by an independent medical researcher and found to have a mean of 22.6 milligrams of nicotine with a standard deviation of 3 milligrams. At level $\alpha = .05$, do the data provide enough evidence to assert that the processor's certification was fallacious?

6 A small factory produces timing devices under subcontract to a major aerospace company. Ideally, the devices have a mean timing advance of .00 second per week. (A *negative* time advance is called a time lag.) A sample of 6 devices, tested as they came off the assembly line, has the following advances:

Serial no. of device	00693A	01014F	01982B	04441W	15672P	18449G
Time advance as tested	.1	.2	−.1	.2	−.1	.0

At significance level $\alpha = .10$, do the data provide evidence that the true mean advance is different from .0, so that the production process requires some adjustment?

7 If a new, slightly cheaper process for mining copper is adopted, it will not be economical if it yields fewer than 50 tons of ore per day in an experimental mine. The results of a 5-day trial period are summarized by the following daily production figures, in tons:

  50   47   53   51   52

At significance level $\alpha = .01$, do the results indicate that the mean production does not fall below 50 tons per day, so that the new process is economical?

SECTION 4.C
TWO-SAMPLE TESTS FOR THE DIFFERENCE OF MEANS

A local board of education, with a view toward eventually cutting costs, conducts a comparison test of the length of time bulbs last before requiring replacement. A sample of 40 randomly selected Universal Electric bulbs, the brand currently used, has a mean lifetime of 1110 hours with a standard deviation of 20 hours. A cheaper brand (Light, Ltd.), which is under consideration as a replacement, is also tested, and a random sample of 60 bulbs turns out to have a mean lifetime of 1081 hours with a standard deviation of 18 hours. The Board of Education would like to use the cheaper bulbs unless the results of the comparison test indicate strongly that the cheaper ones are significantly worse.

LARGE SAMPLES

In answering this question, it is apparent that we are dealing with two distinct samples of data—the 40 Universal Electric bulbs and the 60 Light, Ltd. bulbs. We are comparing these two samples against each other rather than against a hypothesized mean as was done in the previous section. Because more than 30 data points are available, the appropriate statistical test to use is the two-sample z test. If we set

  μ_U = true mean lifetime of *Universal* Electric bulbs

and

  μ_L = true mean lifetime of *Light,* Ltd. bulbs

then the Board of Education wants to test the hypothesis

  H: $\mu_L = \mu_U$ (there is no real difference between the cheaper and the more expensive bulbs) versus

A: $\mu_L < \mu_U$ (the cheaper bulbs are really worse in length of lifetime)

The formula for the two-sample z test for the difference of means is

$$\star \quad z = \frac{\bar{x}_L - \bar{x}_U}{\sqrt{\dfrac{s_L^2}{n_L} + \dfrac{s_U^2}{n_U}}} \tag{4.3}$$

where n_L = number of Light, Ltd. bulbs in sample
$\quad\quad\, n_U$ = number of Universal Electric bulbs in sample
$\quad\quad\, \bar{x}_L$ = sample mean of Light, Ltd. bulbs
$\quad\quad\, \bar{x}_U$ = sample mean of Universal Electric bulbs
$\quad\quad\, s_L$ = sample standard deviation of Light, Ltd. bulbs
$\quad\quad\, s_U$ = sample standard deviation of Universal Electric bulbs

(Here the denominator of z, called the "pooled" standard deviation of the two samples of data, plays the role of $s/\sqrt{n} = \sqrt{s^2/n}$.) By analogy with the second in the list of alternatives presented in Table 4.7, we should reject **H:** $\mu_L = \mu_U$ in favor of **A:** $\mu_L < \mu_U$ at significance level α if $z < -z_\alpha$.

What would be an appropriate level of significance for this problem? Well, because the Light, Ltd. bulbs are cheaper, the Board would prefer to use that brand unless it were convincingly demonstrated that they are really worse. This means that they would not like to reject **H** unless it were really false. Therefore, a test procedure with a very low probability of making a Type I error would be quite appealing to the board. Since the significance level α is the probability of a Type I error, α should be chosen to be very small, say $\alpha = .005$. For this value of α, we would have $z = z_{.005} = 2.57$. It follows that we should reject **H** in favor of **A** if $z < -2.57$.

From the data, we know that

$$n_L = 60 \quad\quad \bar{x}_L = 1081 \quad\quad s_L = 18$$

$$n_U = 40 \quad\quad \bar{x}_U = 1110 \quad\quad s_U = 20$$

Inserting these numbers into their proper places in the formula for z, we get

$$z = \frac{\bar{x}_L - \bar{x}_U}{\sqrt{\dfrac{s_L^2}{n_L} + \dfrac{s_U^2}{n_U}}} = \frac{1081 - 1110}{\sqrt{\dfrac{(18)^2}{60} + \dfrac{(20)^2}{40}}} = \frac{-29}{\sqrt{\dfrac{324}{60} + \dfrac{400}{40}}} = \frac{-29}{\sqrt{5.4 + 10.0}}$$

$$= \frac{-29}{\sqrt{15.4}} = \frac{-29}{3.92} = -7.4$$

We agreed to reject **H** if $z < -2.57$. As it turned out, $z = -7.4 < -2.57$; so we must reject **H**. Our conclusion is that at level $\alpha = .005$ the cheaper Light, Ltd. bulbs are really worse than the Universal Electric bulbs in the sense that they last significantly fewer hours. On the basis of the information gained from this study, we would advise the Board to continue using Universal Electric light bulbs.

SMALL SAMPLES

Example 4.4 Pest Control A state department of public health, faced with a serious infestation of houseflies, decides to conduct a pilot test of effectiveness between two methods of insect control, the organic method and the chemical method. The organic method, which consists of saturating a community with nonpoisonous spiders, is applied in one neighborhood, while the chemical method, using a combination of poisonous sprays and bait, is applied in a second neighborhood comparable to the first. Useful data are collected on six houses in the first neighborhood and on nine in the second during a 48-hour test period. The data are obtained by counting the number of houseflies visually observed making nuisances of themselves during the test period. In the first neighborhood, the one treated by the organic method, the six houses checked had, respectively, 41, 20, 19, 36, 38, and 26 houseflies roaming free. In the second neighborhood, the one treated by the chemical method, the nine houses checked had, respectively 9, 26, 16, 10, 31, 28, 35, 15, and 10 houseflies roaming free. At significance level $\alpha = .01$, do the test results indicate that the chemical method is more effective than the organic method in reducing the number of houseflies roaming free throughout the neighborhood?

SOLUTION We must use the "two-sample t test" because of the relatively small number of data points available, six in one sample and nine in the other. We define the following symbols for the true means relevant to the problem:

μ_o = true mean number of flies successfully evading the *organic* method of insect control

μ_c = true mean number of flies successfully evading the *chemical* method of insect control

We then want to test the hypothesis

H: $\mu_o = \mu_c$ (there is no difference in effectiveness between the organic and chemical methods)

versus

A: $\mu_o > \mu_c$ (the chemical method is more effective)

The number of degrees of freedom df appropriate to the two-sample t test is

$$df = n_o + n_c - 2$$

where n_o = number of data points in first sample (organic)
   n_c = number of data points in second sample (chemical)

In this problem $n_o = 6$ and $n_c = 9$, so that

$$df = n_o + n_c - 2 = 6 + 9 - 2 = 13$$

Because our alternative is **A**: $\mu_o > \mu_c$, a glance at Table 4.7 shows that we should reject **H** in favor of **A** if

$$t > t_\alpha[n_o + n_c - 2] = t_{.01}[13] = 2.650$$

in view of the fact that our significance level here is $\alpha = .01$.

The only task remaining in the solution of this problem is the computation of the numerical value of t, which we have to compare with $t_{.01}[13] = 2.650$. The value of t in the two-sample t test for the difference of means is calculated using the formula

★ $$t = \frac{\bar{x}_o - \bar{x}_c}{\sqrt{\dfrac{\Sigma(x_o - \bar{x}_o)^2 + \Sigma(x_c - \bar{x}_c)^2}{n_o + n_c - 2}\left(\dfrac{1}{n_o} + \dfrac{1}{n_c}\right)}} \qquad (4.4)$$

where $\bar{x}_o$ = sample mean number of houseflies successfully evading organic
   method of control
   $\bar{x}_c$ = sample mean number of houseflies successfully evading chemical
   method of control

The quantities $\Sigma(x_o - \bar{x}_o)^2$ and $\Sigma(x_c - \bar{x}_c)^2$ are the sums of squared deviations from the mean, analogous to the number computed by summing the third column of Table 4.4. We calculate the components of the formula for t by constructing two tables like that of Table 4.4, one such table for each of the two samples of data. The calculations are presented in Table 4.8. From the information presented in Table 4.8, we see that the components of the formula for t are

$$n_o = 6 \qquad \bar{x}_o = 30 \qquad \Sigma(\bar{x}_o - \bar{x}_o)^2 = 458$$
$$n_c = 9 \qquad \bar{x}_c = 20 \qquad \Sigma(\bar{x}_c - \bar{x}_c)^2 = 808$$

Inserting these numbers into their proper places in the formula for t, we obtain

TABLE 4.8

Calculations leading to the two-sample *t* statistic

(Pest control data)

Organic Method of Control			Chemical Method of Control		
x_O	$x_O - \bar{x}_O$	$(x_O - \bar{x}_O)^2$	x_C	$x_C - \bar{x}_C$	$(x_C - \bar{x}_C)^2$
41	11	121	9	−11	121
20	−10	100	26	6	36
19	−11	121	16	−4	16
36	6	36	10	−10	100
38	8	64	31	11	121
26	−4	16	28	8	64
180	0	458	35	15	225
			15	−5	25
			10	−10	100
			180	0	808

$n_O = 6$

$\bar{x}_O = \frac{180}{6} = 30$

$\Sigma(x_O - \bar{x}_O)^2 = 458$

$n_C = 9$

$\bar{x}_C = \frac{180}{9} = 20$

$\Sigma(x_C - \bar{x}_C)^2 = 808$

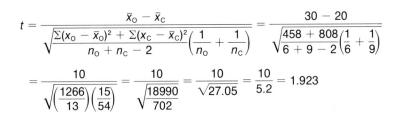

$$t = \frac{\bar{x}_O - \bar{x}_C}{\sqrt{\dfrac{\Sigma(x_O - \bar{x}_O)^2 + \Sigma(x_C - \bar{x}_C)^2}{n_O + n_C - 2}\left(\dfrac{1}{n_O} + \dfrac{1}{n_C}\right)}} = \frac{30 - 20}{\sqrt{\dfrac{458 + 808}{6 + 9 - 2}\left(\dfrac{1}{6} + \dfrac{1}{9}\right)}}$$

$$= \frac{10}{\sqrt{\left(\dfrac{1266}{13}\right)\left(\dfrac{15}{54}\right)}} = \frac{10}{\sqrt{\dfrac{18990}{702}}} = \frac{10}{\sqrt{27.05}} = \frac{10}{5.2} = 1.923$$

We agreed to reject **H** in favor of **A** if *t* turned out to be larger than $t_{.01}$ [13] = 2.650. However, $t = 1.923 < 2.650$, so that we cannot reject **H**. At significance level $\alpha = .01$, we therefore conclude that the data do not provide evidence sufficient to assure us that the chemical method of pest control is more effective than the organic method.

As we pointed out in Chap. 1, one purpose of developing the concepts of the mean and the standard deviation was to make it easier to communicate the basic facts contained in a set of data. In line with this purpose, suppose that only the means and standard deviations of the pest control data were reported to us, instead of the data points themselves. Knowing only that

$n_O = 6$ $n_C = 9$

$\bar{x}_O = 30$ $\bar{x}_C = 20$

$s_O = 9.57$ $s_C = 10.05$

how would we be able to calculate the sums of squared deviations $\Sigma(x_o - \bar{x}_o)^2$ and $\Sigma(x_c - \bar{x}_c)^2$ that are required in formula (4.4)?

Not being in possession of the data points themselves, we are going to have to compute these items indirectly. We use the fact that $\Sigma(x_o - \bar{x}_o)^2$ is mathematically related to s_o by the formula

$$s_o = \sqrt{\frac{\Sigma(x_o - \bar{x}_o)^2}{n_o - 1}}$$

Squaring both sides of the above expression, we see that

$$s_o^2 = \left(\sqrt{\frac{\Sigma(x_o - \bar{x}_o)^2}{n_o - 1}}\right)^2 = \frac{\Sigma(x_o - \bar{x}_o)^2}{n_o - 1}$$

and multiplying through by $n_o - 1$, we obtain

$$\Sigma(x_o - \bar{x}_o)^2 = (n_o - 1)s_o^2$$

A similar calculation shows that

★ $\Sigma(x_c - \bar{x}_c)^2 = (n_c - 1)s_c^2$ (4.5)

and these are the formulas we need. Using them, we find that

$$\Sigma(x_o - \bar{x}_o)^2 = (n_o - 1)s_o^2 = (6 - 1)(9.57)^2 = 458$$

and

$$\Sigma(x_c - \bar{x}_c)^2 = (n_c - 1)s_c^2 = (9 - 1)(10.05)^2 = 808$$

We now are ready to proceed as before to the solution of the problem.

Some examples of situations requiring the use of formula (4.5) are given in the exercises following this section.[1]

Before we go on to another topic, we should point out the conditions under which we may validly use the two-sample t test for the solution of applied problems. Because the table of the t distribution is derived from the table of the normal distribution for the expressed purpose of taking care of small-sample ($n \leq 30$) situations, the two-sample t test should be used only for analyzing normally distributed data. Two further restrictions are required: (1) It must be reasonable for us to assume that both samples of data have approximately the same true standard deviation. (2) Both samples of data must be selected inde-

[1]You figure out which ones they are!

pendently of each other; i.e., individual members of one sample must bear no relation to individual members of the other sample.

Whenever a problem seems to call for the two-sample *t* test, then, the above criteria must be carefully observed so that we can be sure that our solution of the problem is based on sound mathematical principles rather than on the shifting sands of convenience.

EXERCISES 4.C (if the #'s are diff then the problem is Two sample test)

1   A new chemical fertilizer, Nitro-Plus, yielded 20,400 pounds of tomatoes on the average with a standard deviation of 1200 pounds on 40 randomly selected acres of farmland. On another 100 randomly selected acres, the standard organic fertilizer produced a mean yield of 19,000 pounds with a standard deviation of 1000 pounds. At significance level $\alpha = .01$, do the results of the comparison indicate that the chemical fertilizer really produces larger yields than the organic fertilizer?

2   Two methods of teaching remedial reading to 10-year-old youngsters, the Chicago method and the Boston method, were applied to several randomly selected slow readers divided into two groups. On a reading test given to both groups at the conclusion of the teaching program, the 60 pupils ex-posed to the Chicago method scored a mean of 20 with a standard deviation of 7, while the 50 students using the Boston method had an average score of 22 with a standard deviation of 8. At level $\alpha = .05$, can we say that the two methods differ significantly in effectiveness?

3   In a study of whether or not cigarette smoke inhibits the growth of mice, 15 mice born at the same time were fed exactly the same diet in order to prepare them for the experiment. All other living conditions were identical, except that 7 of the mice lived in an atmosphere heavily laden with cigarette smoke, while the other 8 lived in a normal air environment. A biologist participating in the study collects the following data on weight gains of the mice over a period of time:

Weight Gains of Mice, grams	
Atmosphere with Cigarette Smoke	Atmosphere of Normal Air
6	11
7	8
6	10
10	9
8	9
5	6
7	8
	11

At level $\alpha = .05$, are the data sufficient to conclude that cigarette smoke significantly inhibits the growth of mice?

4 Peanuts are known to be an excellent source of protein, and they are therefore a good substitute for meat and fish when prices are high. To find out whether roasted peanuts have less protein than raw peanuts, a nutritionist selected a random sample of 20 bags of peanuts for testing. She chose 8 at random and roasted them and then measured the protein content of all 20 bags. The data follow:

Protein Content of Bags of Peanuts, grams	
Raw Peanuts	Roasted Peanuts
10	8
8	6
6	5
10	10
8	7
10	9
10	6
8	5
6	
10	
4	
6	

At significance level $\alpha = .01$, decide whether or not the roasting process reduces the protein content of peanuts.

5 An economist wants to know whether Commonwealth trade arrangements result in lower prices in Canada for British imports. As an example, he obtains price data on English cheese in six major metropolitan areas in the United States and six in Canada. The data, in cents per pound, follow:

U.S. prices	100	110	130	120	110	90
Canada prices	105	110	100	90	115	105

At significance level $\alpha = .10$, can the economist conclude that English cheese sells at lower prices in Canada?

6 A housing study conducted in 1970 showed that Honolulu had an average monthly rent of 158 dollars per dwelling unit (the highest among U.S. cities) with a standard deviation of 30 dollars, based on the analysis of 12 randomly selected units. In 1975, a random sample of 16 units had a mean rental price of 180 dollars with a standard deviation of 35 dollars. Does the study contain enough information to assert at level $\alpha = .10$ that Honolulu rents increased significantly in the 5-year period?

7 School buildings can be built more economically if state regulations on the required number of windows per classroom are relaxed. In an era of increasing costs combined with an eroding financial structure, the only thing standing between this fact and its immediate implementation is a psychological theory that the lack of windows in the classroom tends to increase anxiety among pupils (presumably because there is no escape). To check out the

theory, an educational psychologist prepared a test of anxiety and administered it to a class of 15 pupils who had spent a month attending classes in an experimental windowless classroom and to a class of 20 pupils (acting as controls) who had spent the same month studying the same material in an ordinary classroom. Those pupils in the windowless classroom had a mean anxiety level of 105 with a standard deviation of 25, while those in the ordinary room had a mean anxiety level of 95 with a standard deviation of 30. At level $\alpha = .01$, do the test results indicate that lack of windows in the classroom leads to a significant increase in anxiety among schoolchildren?

SECTION 4.D
A TEST FOR PAIRED SAMPLES (have to have the same #)

In this section, our goal will be to introduce the "paired-sample t test" which tests the difference between two sets of data in cases where the individual members of one sample are *directly related ("paired")* to corresponding individual members of the other sample. As you can recall from the last two paragraphs of the previous section, the two-sample t test can be used only if the individual members of one sample are independent of the individual members of the other sample. Therefore, the paired-sample t test can be viewed as a replacement for the two-sample t test when the two samples of data are not independent of each other.

Example 4.5 Weight-Reducing Diets A person who feels the need to lose some weight is considering going on one of the two famous reducing diets, the Drink-More diet and the Eat-Less diet, but only if statistical analysis shows that one or both of these diets are effective in inducing weight loss. As it turns out, five acquaintances have tried the Drink-More diet and another five have tried the Eat-Less diet. The weights of the 10 acquaintances, both before beginning and after completing their diet programs, are recorded in Table 4.9. (To avoid embarrassment to those participating in the programs, we have not used their real names.) The question we have to answer is this, "Are either of the diets really effective at significance level $\alpha = .05$ in producing weight loss?"

TABLE 4.9
Before and after weights of ten dieting acquaintances

	Drink-More Diet			Eat-Less Diet	
Acquaintance	Weight Before	Weight After	Acquaintance	Weight Before	Weight After
A	150	150	F	150	141
B	160	160	G	160	150
C	170	170	H	170	160
D	180	179	I	180	170
E	190	141	J	190	179

SOLUTION   As we can discover by looking at the data of Table 4.9, the Drink-More diet is virtually ineffective in the great majority (80%) of cases studied. The Eat-Less diet, on the other hand, seems to let everybody lose approximately 10 pounds. On the surface, it therefore seems that the Eat-Less diet is effective in producing weight loss, but the Drink-More diet is really not effective. This distinction between the diets ought to be reflected in our statistical analysis of the problem.

Recall that the two-sample t test tests whether or not the overall average (true mean) of the group's weight before undertaking the diet was significantly greater than the overall average of the group's weight after completing the diet. In the case of the Drink-More diet, the group weighed a total of 850 pounds (for an average of 170) before dieting and a total of 800 pounds (for an average of 160) after dieting. Therefore, participants in the Drink-More diet lost 10 pounds on the average. The same average results have been obtained in the case of the Eat-Less diet. Participants in that dieting program weighed an average of 170 pounds before and 160 pounds after for an average loss of 10 pounds. It is this *average* loss for which the two-sample t test looks. Because the average weight loss is the same for both diets, the two diets would be equally effective or noneffective in the eyes of the two-sample t test.

But is it the overall average weight loss that we are primarily interested in? We can answer this question by going back to the reasoning that led us to believe that the Drink-More diet was not really effective, but the Eat-Less diet was. As demonstrated in the Drink-More data of Table 4.9, acquaintances A, B, C, and D lost virtually no weight, while acquaintance E lost enough weight for the entire group to have an average loss of 10 pounds, The Eat-Less diet of the same table, on the other hand, shows that each participant individually lost approximately 10 pounds, and so the diet could reasonably be said to guarantee everyone a loss of around 10 pounds. Therefore it was the *individual* changes in weight from before to after, not the overall group averages, that influenced us to think that the Eat-Less diet was effective in producing weight loss, whereas the Drink-More diet was not. It is these individual differences which are not reflected at all by the two-sample t test, for it takes only overall averages into account. Therefore, because the question of the effectiveness of the diets hinges on the individual rather than the overall average weight loss, the two-sample t test is not an appropriate vehicle for answering the question.

What we require in order to make a proper analysis of the diets is a method in which the calculations are based on individual weight losses. Such a method is the paired-sample t test. The paired-sample t test involves not the difference between the average before weight and average after weight, but instead the differences between the individual before and after weights.

We first test the effectiveness of the Drink-More diet using the paired-sample t test. We define the quantity

μ_d = true mean individual difference between the before weight and the after weight and we want to test the hypothesis

H: $\mu_d = 0$ (the diet is not effective)

versus

A: $\mu_d > 0$ (the diet is effective)

The alternative is structured so that the diet is judged to be effective if the average individual loses a positive amount of weight, namely, if the average individual's weight before exceeds his weight after. To carry out the paired-sample t test, we would reject **H** in favor of **A** at level α if $t > t_\alpha[n-1]$, where

★ $\quad t = \dfrac{\bar{d}\sqrt{n}}{s_d}$ $\qquad\qquad\qquad\qquad\qquad\qquad$ (4.6)

where n = number of before-after pairs
$\bar{d}$ = sample mean of individual differences
s_d = sample standard deviation of individual differences

It is usually more convenient, when working with the paired-sample t test, to compute s_d using the shortcut formula for the standard deviation given by

$\quad s_d = \sqrt{\dfrac{n\Sigma d^2 - (\Sigma d)^2}{n(n-1)}}$

as in formula (3.3) of Sec. 3.B.

Here Σd = sum of individual differences
Σd^2 = sum of the squares of individual differences

Taking $\alpha = .05$ as our level of significance and noting that $n = 5$ before-after pairs constitute the "Drink-More" data of Table 4.9, we agree to reject **H** in favor of **A** if $t > t_{.05}[4] = 2.132$. The computations needed for the numerical determination of t are presented in Table 4.10, and they yield that

$\quad t = \dfrac{\bar{d}\sqrt{n}}{s_d} = \dfrac{10\sqrt{5}}{21.81} = \dfrac{(10)(2.236)}{21.81} = 1.025$

Therefore, t has turned out to be 1.025, which is not larger than 2.132; so we cannot reject **H** under the circumstances. We therefore conclude at level $\alpha = .05$, using the paired-sample t test, that the Drink-More diet is not effective in producing individual weight loss.

Now that the paired-sample t test has confirmed our earlier judgment that the Drink-More diet does not produce significant individual weight losses, we await its decision on the Eat-Less diet. It remains to be seen if the paired-sample

TABLE 4.10

Calculations leading to the paired-sample *t* statistic

(Drink-More diet data)

Acquaintance	Weight Before x_B	Weight After x_A	$d = x_B - x_A$	d^2
A	150	150	0	0
B	160	160	0	0
C	170	170	0	0
D	180	179	1	1
E	190	141	49	2401
			Sums 50	2402

$n = 5$

$\bar{d} = \dfrac{\Sigma d}{n} = \dfrac{50}{5} = 10 \ (average \ weight \ loss \ per \ person)$

$s_d = \sqrt{\dfrac{n\Sigma d^2 - (\Sigma d)^2}{n(n-1)}} = \sqrt{\dfrac{5(2402) - (50)^2}{(5)(4)}} = \sqrt{\dfrac{12{,}010 - 2500}{20}}$

$\quad = \sqrt{\tfrac{9510}{20}} = \sqrt{475.5} = 21.81$

t test will be able to detect the substantial individual weight losses that we know are present in the data. We are still testing the hypothesis

H: $\mu_d = 0$

versus

A: $\mu_d > 0$

and we still will reject **H** at level $\alpha = .05$ if $t > 2.132$, because df $= n - 1 = 5 - 1 = 4$, as before. Table 4.11 shows the calculations used in finding the numerical value of *t*. Inserting the numerical values $n = 5$, $\bar{d} = 10$, and $s_d = .7071$ (obtained from the calculations of Table 4.11) into the formula for *t*, we get

$$t = \frac{\bar{d}\sqrt{n}}{s_d} = \frac{10\sqrt{5}}{.7071} = \frac{(10)(2.236)}{.7071} = 31.62$$

We have agreed to reject **H** if $t > 2.312$. As it turns out, $t = 31.62$, a number substantially in excess of 2.312, so that we can proceed with the greatest confidence to the rejection of **H**. In the eyes of the paired-sample *t* test, then, there is little doubt that the Eat-Less diet is effective at level $\alpha = .05$ or any other level for that matter. The numerical value $t = 31.62$ is so overwhelming that we must conclude, at any reasonable level α, that the Eat-Less diet is truly effective in producing individual weight loss.

TABLE 4.11

Calculations leading to the paired-sample t statistic
(Eat-Less diet data)

Acquaintance	Weight Before x_B	Weight After x_A	$d = x_B - x_A$	d^2
F	150	141	9	81
G	160	150	10	100
H	170	160	10	100
I	180	170	10	100
J	190	179	11	121
			Sums 50	502

$n = 5$

$$\bar{d} = \frac{\Sigma d}{n} = \frac{50}{5} = 10$$

$$s_d = \sqrt{\frac{n\Sigma d^2 - (\Sigma d)^2}{n(n-1)}} = \sqrt{\frac{5(502) - (50)^2}{(5)(4)}} = \sqrt{\frac{2510 - 2500}{20}}$$

$$= \sqrt{\tfrac{10}{20}} = \sqrt{.5} = .7071$$

EXERCISES 4.D

1 Some psychologists feel that there is a statistical correlation between smoking and absenteeism. The management of a leather-goods factory has under consideration a stop-smoking incentive plan, and they would therefore be interested in knowing whether employees who have stopped smoking have better absenteeism records than they had while they were smoking. Nine such employees are randomly selected, and their absenteeism records before and after they have stopped smoking are compared. The data follow:

Employee	A	B	C	D	E	F	G	H	I
Days absent per year while smoking	20	30	14	6	42	19	18	12	24
Days absent per year after stopping smoking	10	20	16	5	40	15	22	10	20

At significance level $\alpha = .05$, does the comparison tend to support the theory that employees have less absenteeism after they stop smoking?

2 At various points along a valley, a team of geographers measured the angles of slope on both sides of the valley wall. Their problem was to determine whether or not there were significant differences between the slope angles of the northeast (NE) slope and the southwest (SW) slope. The data they collected are as follows:

Miles Upstream	10	20	30	40	50	60	70	80
NE slope, degrees	15	24	28	31	35	34	35	36
SW slope, degrees	16	22	25	30	34	32	36	34

At level $\alpha = .05$, can the researchers conclude that the angles of slope of the northeast wall differ significantly from the angles of slope of the southwest wall?

3 Ten sets of identical twins were administered drugs intended to increase the pulse rate. The goal of the experiment was to determine if identical twins have similar reactions to medication. The following data on increase of pulse rates was observed:

Twin Set	Pulse Rate Increase of Older Twin	Pulse Rate Increase of Younger Twin
#1	12	19
#2	14	13
#3	8	6
#4	11	18
#5	14	12
#6	12	15
#7	13	10
#8	15	18
#9	18	21
#10	17	22

At level $\alpha = .10$, decide whether or not identical twins react similarly to medication affecting the pulse rate.

4 Environmental Nucleonics of Barstow instituted a time-consuming and rather expensive occupational health and safety program, the goal of which was to reduce the number of hours of working time lost due to on-the-job accidents. The following data were collected in their six branches before and after institution of the program:

Hours Lost Per Employee Per Quarter		
Branch	Before	After
Barstow, Ca.	46	43
Hoboken, N.J.	72	66
Ensenada, B.C.	50	51
Elko, Ne.	38	37
Rawlins, Wy.	78	73
Vancouver, B.C.	66	60

At level $\alpha = .01$, decide whether or not the data indicate that the program is effective in reducing time lost due to on-the-job accidents.

5 An agricultural experiment station tested the comparative yields of two vari-

eties of corn, a new experimental variety and a standard variety, to find out whether or not the new variety would permit greater food production on the same land. Each of seven farmers was asked to grow both varieties on similar plots of land and to give each the same amount of treatment, water, fertilizer, etc. The yields, in bushels, are recorded below:

Farmer	A	B	C	D	E	F	G
New variety	28.2	24.6	29.7	20.5	34.6	27.1	31.4
Standard variety	27.2	24.3	29.0	22.5	34.2	26.8	30.4

At level $\alpha = .05$, do the results of the experiment confirm the claim that the new variety is significantly more productive than the standard variety?

6 A local school system paired 16 girls in the first grade of elementary school according to psychological test scores, socioeconomic status of family, general health, and family size, as part of a study of the effectiveness of the kindergarten experience in helping a pupil in first grade. One member of each pair had attended an optional kindergarten the year before, while the other had not. Halfway through the year (first grade) the 16 girls were tested for mastery of the first-grade material taught, and they had the following scores:

Pair	#1	#2	#3	#4	#5	#6	#7	#8
With kindergarten	83	74	67	64	70	67	81	64
Without kindergarten	73	74	63	66	68	63	77	65

Does the study indicate, at level $\alpha = .05$, that the kindergarten experience seems to help first graders?

SECTION 4.E
TESTS FOR PROPORTIONS

As has been pointed out in Secs. 2.B and 3.C, many questions involving proportions can be better understood if they are structured according to the format of the binomial distribution. The large-sample tests of proportions are based on percentage points of the normal distribution because for large numbers of data points the normal approximation to the binomial distribution comes into play.

The following example illustrates the "one-sample z test for the proportion of a population."

Example 4.6 A Political Poll A local journalist reports, after talking to all his friends, that 60% of the state's electorate supports the reelection of the incumbent U.S. senator. The challenger naturally takes issue with this report and orders his political science research staff to check the accuracy of the claim. The staff polls 1500 voters statewide and finds that 784 of those polled support the incumbent. At level $\alpha = .01$, does the poll tend to refute the journalist's report?

SOLUTION   Denoting by p the true proportion of voters who support the incumbent, we want to test the hypothesis

H: $p = .60$ (the true proportion supporting the incumbent is 60%)

against the alternative

A: $p < .60$ (less than 60% support the incumbent)

In accordance with the principles of the normal approximation to the binomial distribution (since $n = 1500 > 30$), we compute

★ $$z = \frac{\hat{p} - p_0}{\sqrt{\dfrac{p_0(1 - p_0)}{n}}}$$ (4.7)

where p_0 = hypothesized proportion supporting incumbent
$\hat{p}$ = sample proportion supporting incumbent
n = number of voters involved in poll

(It should be noted that the formula is reminiscent of the one used in the one-sample z test for the mean of a population.) We then reject **H** at level α if $z < -z_\alpha$. This rejection rule has been chosen on the basis of the three options presented in Table 4.7, upon replacing the μ's by p's.

We therefore agree to reject **H:** $p = .60$ in favor of **A:** $p < .60$ at level $\alpha = .01$ if

$z < -z_{.01} = -2.33$

It now remains only to compute the numerical value of z. From the data accumulated in the poll, we know that

$$n = 1500 \quad \text{and} \quad \hat{p} = \frac{784}{1500} = .5227$$

and, on the basis of the question asked, we see that the hypothesized proportion is

$p_0 = .60$

It follows automatically that

$$z = \frac{\hat{p} - p_0}{\sqrt{\dfrac{p_0(1 - p_0)}{n}}} = \frac{.5227 - .60}{\sqrt{\dfrac{(.60)(.40)}{1500}}} = \frac{-.0773}{\sqrt{.00016}} = \frac{-.0773}{.01265} < -6.11$$

As we agreed to reject **H** if z turned out to be less than (more negative than!) −2.33, the result that $z = -6.11 < -2.33$ leads us to reject **H**. At level $\alpha = .01$, the challenger's poll therefore tends to refute the journalist's report that 60% of the voters support the incumbent. The polls indicate that the true percentage supporting the incumbent is significantly below 60%.

The following situation calls for the "two-sample z test for the difference of proportions."

Example 4.7 Advertising Effectiveness The advertising department of a major Southern California automobile dealership which has branches in Orange and Los Angeles Counties wants to know whether its advertising is really effective in influencing prospective customers. In particular, the department wants to find out if more station wagons can be sold by saturating an area with advertising urging customers to buy a station wagon. To test out the effectiveness of such an advertising campaign, the dealership floods Orange County for 1 week with advertising while maintaining a low profile in Los Angeles County. Over the next 2 months 120 of 400 cars it sells at the Orange County branch are station wagons, while 150 of 600 cars it sells at the Los Angeles branch are station wagons. Do the results of the study tend to indicate that a proportionately higher number of station wagons are sold in Orange County where the advertising was concentrated? Use a level of significance of $\alpha = .05$.

SOLUTION   We define some symbols as follows:

p_H = true proportion of station wagons sold in region of *heavy* advertising
p_N = true proportion of station wagons sold in region of *normal* advertising

We want to test the hypothesis

H: $p_H = p_N$ (advertising does not affect sales of station wagons)

against the alternative

A: $p_H > p_N$ (heavy advertising increases sales of station wagons)

n_H = number of data points in region of heavy advertising
$\hat{p}_H$ = sample proportion of station wagons sold in region of heavy advertising
n_N = number of data points in region of normal advertising
$\hat{p}_N$ = sample proportion of station wagons sold in region of normal advertising
$\hat{p}$ = sample proportion of station wagons sold in both regions combined

Sales records indicate that

$$n_H = 400 \qquad\qquad\qquad n_N = 600$$

$$\hat{p}_H = \frac{120}{400} = .30 \qquad\qquad\qquad \hat{p}_N = \frac{150}{600} = .25$$

$$\hat{p} = \frac{120 + 150}{400 + 600} = \frac{270}{1000} = .27$$

In accordance with the rejection rules of Table 4.7, we calculate

$$\bigstar \quad z = \frac{\hat{p}_H - \hat{p}_N}{\sqrt{\hat{p}(1 - \hat{p})\left(\dfrac{1}{n_H} + \dfrac{1}{n_N}\right)}} \qquad\qquad (4.8)$$

H = heavy advertising
N = Normal advertising

(The formula might remind you of the one for the two-sample t test for the difference of means.) Then we reject **H**: $p_H = p_N$ in favor of **A**: $p_H > p_N$ at level α if $z > z_\alpha$. Here $\alpha = .05$; so we agree to reject **H** if

$$z > z_{.05} = 1.64$$

Upon inserting the numerical values of the components of the formula for z, we get

$$z = \frac{\hat{p}_H - \hat{p}_N}{\sqrt{\hat{p}(1 - \hat{p})\left(\dfrac{1}{n_H} + \dfrac{1}{n_N}\right)}} = \frac{.30 - .25}{\sqrt{(.27)(.73)\left(\dfrac{1}{400} + \dfrac{1}{600}\right)}}$$

$$= \frac{.05}{\sqrt{(.197)\left(\dfrac{1000}{240,000}\right)}} = \frac{.05}{\sqrt{(.197)(.004167)}} = \frac{.05}{\sqrt{.00082}}$$

$$= \frac{.05}{.0286} = 1.75$$

Therefore $z = 1.75 > 1.64$; so we have to reject **H** in favor of **A** at level $\alpha = .05$. We conclude that at level $\alpha = .05$, the advertising campaign can be considered to have been effective in increasing the proportion of station wagons sold.

One final remark on the previous example: If the proper level of significance had been $\alpha = .01$, in the opinion of the advertising department, the advertising campaign would have been judged noneffective. This is due to the fact that the rejection rule would have indicated the rejection of **H** when $z > z_{.01} = 2.33$. Since z turned out to be 1.75, **H** could not validly be rejected and we would have to say that the advertising did not seem to affect sales. We have here one more example of the importance of making an appropriate subjective choice of the significance level α, the probability of rejecting **H** when it's really true.

EXERCISES 4.E

1   To find out whether or not a particular coin is really "fair" (equally likely to fall heads or tails), the coin is tossed several times and the number of heads resulting is carefully noted.
   (a)   If 40 out of 100 tosses result in heads, can we assert at level $\alpha = .01$ that the coin is fair?
   (b)   If 400 out of 1000 tosses result in heads, can we conclude at level $\alpha = .01$ that the coin is fair?

2   An executive in charge of new-product development at a local dairy feels that at least 10% of the population would be willing to try a carton of chocolate flavored buttermilk, while the other 90% would not be at all interested. Is her opinion substantiated at level $\alpha = .10$ by a survey of 200 randomly selected persons of which 16 expressed interest?

3   The manufacturer of a new chemical fertilizer advertises that with the aid of his fertilizer, at least 80% of the seeds planted will germinate. Does an experiment in which 130 seeds out of 170 germinate tend to refute his claims at level $\alpha = .05$?

4   A "straw poll" of 140 randomly selected state legislators turns up 70 who favor a certain controversial change in the banking laws.
   (a)   At level $\alpha = .10$, do the results of the poll conflict with a news reporter's assertion that at least 60% of the legislators support the change?
   (b)   At level $\alpha = .10$, do the results of the poll conflict with a banking lobbyist's assertion that no more than 40% of the legislators support the change?

5   One public health physician conducts a study to determine whether persons in his geographical area of responsibility are more likely to have lung disorders if they are heavy smokers than if they are not. Of 200 heavy smokers aged 50 or over, 10% report some sort of lung problem, while of 300 persons who are not heavy smokers, only 7% suffer from lung ailments. At level $\alpha = .05$, can the physician conclude that heavy smokers are significantly more likely than other persons to contract lung disorders?

6   A survey of 50 licensed drivers living in an urban area turns up 8 who have a drinking problem, while a survey of 30 drivers in a rural area reveals that 4 are problem drinkers. Can we conclude at level $\alpha = .05$ that the proportion of problem drinkers differs significantly between urban and rural areas?

SUMMARY AND DISCUSSION

In this chapter, we have introduced the procedure of formulating a question in statistical terms and then answering it using available data. As we have seen, the first step is setting up a hypothesis and an alternative, and the second is using the data to help us choose between them. We have illustrated the technique of testing statistical hypotheses in examples involving means and proportions. The general problem of statistical inference that we first encountered in our study of confidence intervals in Chap. 3, namely, the fact that we cannot be 100% sure of our decision, is still with us. It is reflected in the fact that each statistical decision is made in terms of a level of significance, which is our probability of rejecting the hypothesis we formulated when that hypothesis is actually true. Having chosen a level of significance gives us an idea of our chances of making an incorrect choice between hypothesis and alternative.

SUPPLEMENTARY EXERCISES

1   The manufacturer of a hospital room humidifier advertises that the control dial on the device will maintain a mean room humidity of 80 with a standard deviation of 2.0. Its performance was carefully observed during a 1-day test at 60 different times throughout the day. The mean humidity level turned out to be 78.3 with a standard deviation of 2.9. At level $\alpha = .01$, do the observed data tend to contradict the manufacturer's claim that the true mean humidity level will be 80?

2   In a routine check of drug packaging, the FDA counts the contents of 16 bottles of aspirin labeled as containing 500 tablets each. The 16 bottles have mean contents 494 aspirins with a standard deviation of 30. Can the FDA assert at level $\alpha = .05$ that the bottles are underfilled?

3   A company which markets canned carrots advertises that their 15-ounce cans actually have mean weight 15.1 ounces with a standard deviation of .1 ounce, so that a substantial percentage of cans exceeds the stated contents (15.0 ounces). To test the validity of the company's assertion, a state consumer agency selects a random sample of 11 cans and carefully weighs the contents of each of them. The 11 weights are as follows:

15.1	15.0	15.2	14.9	15.0	15.1	15.1	14.9
15.1	15.0	14.6					

(a)   At significance level $\alpha = .05$, can the agency assert that the true mean contents are below the 15.1 ounces claimed by the company?

(b)   At level $\alpha = .05$, do the data indicate that the true mean contents are below the 15.0 ounces printed on the label?

4   A truck farmer decides to test a new nonpolluting insecticide which also, according to the inventor, definitely reduces the loss attributable to a common pest. The farmer has a feeling that the new spray is really worse than the standard (polluting) spray; so he treats 31 acres with the new spray and 41 with the standard spray throughout the growing season. For the new insecticide, the mean yield per acre turns out to be 980 pounds with a standard deviation of 60 pounds, while for the standard spray there is a mean yield of 1040 pounds with a standard deviation of 50 pounds. At level $\alpha = .01$, do the results of the test provide evidence that the new spray is significantly worse than the standard one?

5 In order to compare the merits of two short-range ground-to-ground an-titank weapons, 8 of the conventional type and 10 of the more expensive wire-guided rockets are tested. The conventional type have mean target error of 5.2 feet with a standard deviation of 1.8 feet, while the wire-guided ones have mean target error of 3.6 feet with a standard deviation of 1.5 feet. At level $\alpha = .10$, can we say that the more expensive rockets are really more accurate?

6 After the recent detonation of a 20-kiloton nuclear bomb in the atmosphere at the Lop Nor testing site in northwestern China, scientists in Japan and the United States immediately began to measure the increase in radiation levels due to the nuclear fallout. The radiation cloud, released directly into the atmosphere, was first detected by instruments in Japan 5 days after the test, and it reached eight western states of the United States a week later. The following data give the radiation amounts (in picocuries of radium per cubic meter of air, above the usual natural level of 2.1) at six geographic stations in Japan and eight in the United States:

Japan	.2	.4	.5	.1	.3	.3		
United States	.2	.3	.1	.1	.2	.2	.4	.1

At level $\alpha = .05$, decide whether nuclear fallout levels in Japan are greater than those in the United States after release of a radiation cloud into the atmosphere over China.

7 The research department of a company which produces industrial string proposes a new technique for strengthening the string and making it able to withstand stronger forces. To find out whether the new technique really does strengthen the string, five lengths of string are produced by the stan-dard technique and five by the new technique, and the samples are com-pared for their respective breaking points, in pounds of force. The breaking points are as follows:

Standard technique	144	131	155	126	134
New technique	139	154	132	143	147

At level $\alpha = .10$, does the comparison support the contention of the re-search department that the new technique produces stronger string?

8 Several insurance adjusters were concerned about the unusually high re-pair estimates they seemed to be getting from Fosbert's U-bet Repair Sta-tion. To test their suspicions, they brought each of 8 damaged cars to Fosbert's and also to the auto-repair shop of Nickle's Department Store, a concern generally regarded as reliable. They obtained the following esti-mates, in hundreds of dollars:

Car Number	#1	#2	#3	#4	#5	#6	#7	#8
Fosbert's estimate	2.1	4.5	6.3	3.0	1.2	5.4	7.3	9.3
Nickle's estimate	2.0	3.8	5.9	2.8	1.3	5.0	6.5	8.6

At level $\alpha = .01$, can the adjusters conclude from their survey that Fosbert's estimates are significantly higher than Nickle's?

9   A committee of elementary-school teachers is assigned the task of comparing the new math with the old math in the ability to develop pupils' skills in arithmetic. Twenty pupils (10 from a group being taught the new math and 10 from a group being taught the old math) were paired on the basis of age, athletic ability, family income, and general health, and then the committee administered the same arithmetic skills test to all 20 students. The scores follow:

Pair	A	B	C	D	E	F	G	H	I	J
New math group	100	50	10	60	80	60	80	40	20	20
Old math group	90	60	20	60	70	70	90	50	10	40

Do the test scores indicate at level $\alpha = .05$ that the committee should report that the new math is less effective than the old math in developing arithmetic skills?

10   A compendium of actuarial tables published in 1958 shows that of all 45-year-old men in a certain occupational grouping, 7% die before reaching age 55. A more recent survey of 2000 men in the age and occupational group reveals that percentage to be 5%. At level $\alpha = .01$, does the recent survey indicate that the probability of death in that group has been significantly reduced since 1958?

11   One urban affairs sociologist claims that 55% of the adult population of a particular major city have been victimized by a criminal at one time or another.
   (a)   Does a random sample of 180 residents of whom 110 have been victimized tend to refute his claim at level of significance $\alpha = .05$?
   (b)   Does a random sample of 160 residents of whom 80 have been victimized tend to refute the claim at level $\alpha = .05$?

12   As part of an analysis of overdue accounts which eventually require legal action to force payment, a corporation accountant discovers that 45 out of 130 corporate accounts required legal action in the past, while 214 of 1070 individual accounts required such action. Can the accountant conclude that a corporate account is significantly more likely to require legal action than is an individual account? Use level $\alpha = .01$.

13   A political poll concerning the election in which Able is running against Baker shows that 520 of 1000 urban voters favor Able, while 240 of 500 rural voters favor Able. Can we conclude from the results of the poll that Able is, at level $\alpha = .05$, significantly more popular in urban areas than in rural areas?

REGRESSION AND CORRELATION

In a study of the efficiency of automobile engines with respect to consumption of gasoline, one particular 3990-pound car was operated at various speeds, and the gasoline consumption (in miles per gallon) at each speed level was carefully measured. The data of Table 5.1 expresses the results of the experiment. Looking at the data, we can immediately draw one simple conclusion: at higher speeds, there is less efficient use of gasoline; namely, the car gets fewer miles per gallon of gas. The techniques of regression and correlation were developed to answer the following four more specific questions:

1   Is there a simple mathematical relationship (i.e., a relationship expressible by a simple formula) between speed, in miles per hour, and efficiency, in miles per gallon?
2   If so, what is this relationship?
3   How much of the variation in efficiency level can be attributed to variations in

TABLE 5.1
Efficiency of an automobile engine

Speed, miles/hour	Efficiency, miles/gallon
30	20
40	18
50	17
60	14
70	11

speed; i.e., among all those factors which tend to affect gasoline efficiency, how important a role is played by the speed of the car?

4 Can we use the data to predict the efficiency at other speeds, for example, 25, 55, or 80?

SECTION 5.A
THE GEOMETRY OF THE STRAIGHT LINE

The method of illustration of pairs of data used to visualize a relationship is called a "scattergram." The data of Table 5.1 are illustrated by the scattergram of Fig. 5.1, composed of a horizontal axis on which speed levels are marked and

FIGURE 5.1
Scattergram of speed versus efficiency.

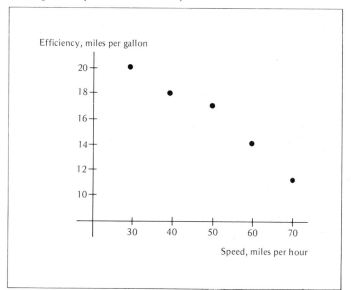

a vertical axis on which efficiency levels are indicated. Each trial of the experiment is represented by a single dot indicating both the speed and the efficiency levels. For example, the dot for 50 miles/hour is 50 units to the right of the vertical axis and 17 units above the horizontal axis because the corresponding efficiency level was 17 miles/gallon.

The scattergram of Fig. 5.1 communicates to the viewer information beyond that contained in the earlier statement that, at higher speeds, there is less efficient use of gasoline. From the scattergram, we can see that there seems to be a definite linear trend which, while not perfect, does describe the general characteristics of the relationship between speed and efficiency. Figure 5.2 shows three possible candidates for the straight line which best fits the data. We would like to select from among all the possible straight lines the one best-fitting line.

Every straight line has its own equation, which is a mathematical description of the line's slope and location. For example, let's consider the line which joins the two points in Fig. 5.3. One of the points is labeled (1, 2) to indicate that it lies at a perpendicular distance 1 to the right of the vertical axis and a perpendicular distance 2 above the horizontal axis. The other is labeled (5, 4) because it lies 5 units to the right of the vertical axis and 4 units above the horizontal axis. [Note: A point labeled (−4, −3) would lie 4 units to the *left* of the vertical axis and 3 units *below* the horizontal axis.]

The line joining (1, 2) with (5, 4) also contains several other points—infinitely many, to be sure. In fact, a line can be viewed as a collection of points, all "glued" together in a special configuration. If a typical point is labeled (x, y)

FIGURE 5.2
Possible best-fitting lines.

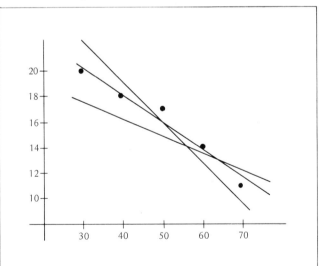

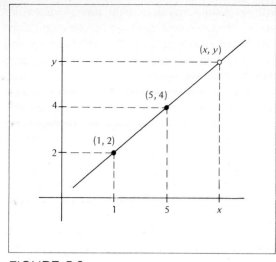

FIGURE 5.3
A typical straight line.

when it lies *x* units to the right of the vertical axis and *y* units above the horizontal axis, some of these points (*x*, *y*) happen to fall exactly on the line, and some do not. Actually, relatively few of the points (*x*, *y*) do fall on the line in question—most do not. When (*x*, *y*) does fall on the line, two triangles are formed, as in Fig. 5.4, a smaller triangle inside a larger triangle. When (*x*, *y*) falls

FIGURE 5.4
(*x*, *y*) on the line.

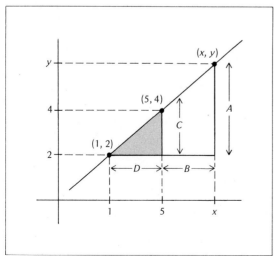

off the line, no such "nested" triangles are formed, but a situation like that in Fig. 5.5 occurs instead.

In Fig. 5.4, where (x, y) falls on the line joining $(1, 2)$ with $(5, 4)$, the triangles formed are similar. "Similar to what?," you may ask. Well, you might recall from plane geometry that two triangles are said to be similar if they have the same shape, although they need not be the same size. In such triangles, corresponding sides are proportional; for example, in Fig. 5.4, we know that A is to $B + D$ as C is to D. Algebraically, this means that (x, y) lies on the line if and only if[1]

$$\frac{A}{B + D} = \frac{C}{D}$$

Now the geographic layout of Fig. 5.4 indicates that $A = y - 2$, being the distance from 2 to y on the vertical axis. Similarly, $B + D = x - 1$, $C = 4 - 2 = 2$, and $D = 5 - 1 = 4$. Therefore, making these substitutions we get

$$\frac{y - 2}{x - 1} = \frac{2}{4}$$

Using cross-multiplication and other algebraic techniques, we have that (x, y)

[1] For example, suppose $x = 9$ and $y = 6$. Then $A = 4$, $B = 4$, $C = 2$, and $D = 4$. Substituting, we get $4/(4 + 4) = 2/4$, which is true, so that $(9, 6)$ lies on the line. On the other hand, suppose $x = 10$ and $y = 7$. Then $A = 5$, $B = 5$, $C = 2$, and $D = 4$. Substituting, we get $5/(5 + 4) = 2/4$, which is not true; so $(10, 7)$ does not lie on the line.

FIGURE 5.5
(x,y) off the line.

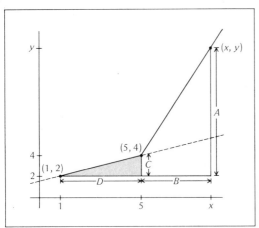

lies on the line if and only if, in sequence.

$$4(y - 2) = 2(x - 1)$$
$$4y - 8 = 2x - 2$$
$$4y = 2x + 6$$

and finally

$$y = \tfrac{1}{4}(2x + 6)$$
$$= .5x + 1.5$$
$$= 1.5 + .5x$$

This means that (x, y) lies on a line joining $(1, 2)$ with $(5, 4)$ if and only if

$$y = 1.5 + .5x$$

That equation is called "the equation of the line joining $(1, 2)$ with $(5, 4)$." The equation specifies which points (x, y) are on the line and which are not, for it says that (x, y) is on the line if and only if y is 1.5 more than half of x. For example, $(5, 4)$ is on the line because $4 = 1.5 + (.5)(5)$, $(9, 6)$ is on the line because $6 = 1.5 + (.5)(9)$, and $(11, 8)$ is not on the line because $8 \neq 1.5 + (.5)(11)$.

In the same way, we could start with any two points and work out the equation of the line joining them. Because the geometric and algebraic techniques used would be the same, we would come up with an equation of the form

$$y = a + bx$$

where a and b are numbers. Therefore every straight line has an equation of the form $y = a + bx$.

Conversely, from any equation of the form $y = a + bx$, we can draw the corresponding straight line. For example, consider the equation

$$y = -2 + 3x$$

If we can find any two points on the line represented by this equation, we can connect them in order to obtain the graph of the line. We can find points by picking x values at random and then working out the corresponding y values. If we choose $x = 0$, then $y = -2 + (3)(0) = -2 + 0 = -2$; so the point $(0, -2)$ is on the line. If next we select $x = 2$, then $y = -2 + (3)(2) = -2 + 6 = 4$, so that $(2, 4)$ lies on the line. Connecting the points $(0, -2)$ and $(2, 4)$, as in Fig. 5.6, we get a picture of the line. Notice that an individual choosing $x = 1$ and $x = 3$ instead of

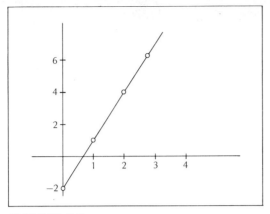

FIGURE 5.6
The line $y = -2 + 3x$.

$x = 0$ and $x = 2$ would obtain the same line by use of the points (1, 1) and (3, 7).
All these points lie on the line whose equation is $y = -2 + 3x$.

EXERCISES 5.A

1 For each of the following pairs of points, determine the equation of the
  straight line passing through them:
  (a) (1, 2) and (5, 3)
  (b) (0, −2) and (2, 1)
  (c) (1, 9) and (5, 5)
2 For each of the following pairs of points, determine the equation of the
  straight line passing through them:
  (a) (1, 2) and (5, 6)
  (b) (0, −2) and (2, 5)
  (c) (1, 9) and (5, 8)
3 For each of the following equations, find two points on the corresponding
  straight line and draw the line:
  (a) $y = 6 + 2x$.
  (b) $y = 8 - 3x$.
  (c) $y = -3 + 4x$.
4 For each of the following equations, find two points on the corresponding
  straight line and draw the line:
  (a) $y = 8 + 3x$.
  (b) $y = 6 - 2x$.
  (c) $y = -4 + 3x$.
5 Determine the equation of the straight line passing through the points (8.1,
  2.3) and (3.2, 1.8), and find a third point lying on that line.
6 Find two points lying on the straight line having equation $y = 7.8 - 2.2x$, and
  draw the line.

SECTION 5.B
THE REGRESSION LINE AND PREDICTION

The line which best fits the points of Fig. 5.1 is called the regression line of the points. We know from Sec. 5.A that the regression line has equation $y = a + bx$, where a and b are particular numbers. Therefore, in order to find the equation of the regression line, we only have to determine the numerical values of a and b.

Let's first specify exactly what we mean by the phrase "best-fitting line," one that would best fit the purposes for which we want to eventually use it. The data of Table 5.1 relate the gasoline efficiency of the engine to the speed of the car. The regression line would be a mathematical relationship between speed and efficiency, and we would most likely use such a relationship to predict efficiency at various speeds. It is traditional, algebraically speaking, to label speed by x, the "independent" variable, and efficiency by y, the "dependent" variable. Then the equation $y = a + bx$ expresses the manner in which efficiency "depends on" speed. A statement such as this does not necessarily imply any causal dependence, like a cause-and-effect relationship, but merely an algebraic connection between the numbers representing efficiency and the numbers representing speed. We agree then to label efficiency as y and speed as x, and, in general, we shall always use y to denote the variable to be predicted.

A prediction of efficiency based on the regression line $y = a + bx$ will be as accurate as possible only if we choose a and b in such a way that the vertical distances between the data points and the line (known as "vertical deviations") are collectively as small as possible. This situation is illustrated in Fig. 5.7, where the original data points are compared with the location predicted for them by a possible regression line $y = a + bx$. By analogy with the development of the mean and standard deviation in Chap. 1, we consider the regression line as a sort of mean or balancing line of the data points. Since the positive and negative vertical deviations of the data points from the line should cancel each other, the real deviation of the points from the line will have to be calculated using squares in order to eliminate the cancellation effects of points falling above and below the line.

A typical set of n data pairs $(x_1, y_1), (x_2, y_2), \ldots, (x_n, y_n)$ to be fitted to a line $y = a + bx$ will have its sum of squared deviations equal to

$$SSD = \Sigma(y - a - bx)^2$$

For a particular set of data points, the set given in Table 5.1 for example, a line having equation $y = a + bx$ wil be the best-fitting line if we use the values of a and b which result in the *smallest* possible value of SSD, the sum of squared deviations. The regression line resulting from this method of reasoning is often called "the least-squares line."

For different sets of data points, we would expect to come up with different numerical values of a and b, because the regression lines would most likely be

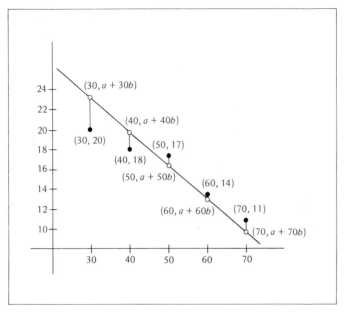

FIGURE 5.7

Vertical deviations of data points from the line $y = a + bx$.

different in each case. So what we really need are formulas for a and b that we can use to compute their numerical values from the data. The numbers a and b can be determined from the data using the formulas

$$\bigstar \quad b = \frac{n\Sigma xy - (\Sigma x)(\Sigma y)}{n\Sigma x^2 - (\Sigma x)^2} \qquad (5.1)$$

and

$$\bigstar \quad a = \frac{\Sigma y - b\Sigma x}{n} \qquad (5.2)$$

It should be noted that the numerical value of b must be calculated first, because it is itself used in the calculation of a. Another important observation is that Σx^2 is not the same as $(\Sigma x)^2$. For example, if $x_1 = 2$, $x_2 = 5$, and $x_3 = 1$, we would have

$$\Sigma x^2 = 2^2 + 5^2 + 1^2 = 4 + 25 + 1 = 30$$

but

$$(\Sigma x)^2 = (2 + 5 + 1)^2 = 8^2 = 64$$

The formulas for *a* and *b* involve sums of several quantities, namely, the following:

Σx = sum of the *x* values of the data points

Σy = sum of the *y* values of the data points

Σx^2 = sum of the squares of the *x* values

Σxy = sum of the products of each *x* multiplied by its corresponding *y*

The simplest way to keep track of all the calculations necessary to obtain the regression line is by way of a table of the sort illustrated in Table 5.2. (In our table, we will also include the calculation of Σy^2, the sum of squares of the *y* values, which will turn out to be needed in measuring how useful the regression line will be in predicting future occurrences.) Using the sums obtained in Table 5.2, we can complete the calculation of *a* and *b* by inserting the sums into their proper places in the formulas. This is done as follows:

$$b = \frac{n\Sigma xy - (\Sigma x)(\Sigma y)}{n\Sigma x^2 - (\Sigma x)^2} = \frac{5(3780) - (250)(80)}{5(13,500) - (250)^2}$$

$$= \frac{18,900 - 20,000}{67,500 - 62,500} = \frac{-1100}{5000} = -.22$$

$$a = \frac{\Sigma y - b\Sigma x}{n} = \frac{80 - (-.22)(250)}{5} = \frac{80 + 55}{5} = 27$$

Having calculated $a = 27$ and $b = -.22$, we therefore know that the equation of the regression line

$$y = a + bx$$

TABLE 5.2
Preliminary calculations in linear regression

Speed x, miles/hour	Efficiency y, miles/gallon	x^2	y^2	xy
30	20	900	400	600
40	18	1600	324	720
50	17	2500	289	850
60	14	3600	196	840
70	11	4900	121	770
Sums $n = 5$ 250	80	13,500	1330	3780

is, in this case,

$$y = 27 - .22x$$

It is often useful, for purposes of visual communication and analysis, to superimpose the graph of the regression line upon the scattergram of the data. The scattergram for the engine efficiency problem appears in Fig. 5.1. We superimpose the graph of the line

$$y = 27 - .22x.$$

on the scattergram by choosing two points on the line and connecting them as in the discussion relating to Fig. 5.6. If we choose $x = 30$, then $y = 27 - .22x = 27 - (.22)(30) = 20.4$, so that (30, 20.4) is on the regression lines. Choosing $x = 60$, the corresponding $y = 27 - (.22)(60) = 13.8$, so that (60, 13.8) is a second point on the regression line. The two points (30, 20.4) and (60, 13.8) are graphed together with the regression line in Fig. 5.8.

We are now ready to tackle the fourth question raised at the beginning of the chapter. Can we use the data to predict the efficiency at other speeds, for example 25, 55, or 80?

We first look at the regression line relating speed x and engine efficiency y which has equation $y = 27 - .22x$. In Fig. 5.8, we have represented this relationship graphically in a manner that illustrates the linear trend of variation in en-

FIGURE 5.8
Superposition of regression line upon scattergram.

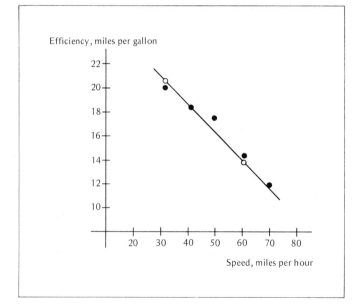

TABLE 5.3
Predicted efficiency levels for various speeds

Speed x, miles/hour	Computation $27 - .22x$	Predicted Efficiency y, miles/gallon
25	$27 - (.22)(25)$	21.5
55	$27 - (.22)(55)$	14.9
80	$27 - (.22)(80)$	9.4

gine efficiency relative to variations in speed. What this means is that, insofar as we are able to determine from the five data points, when the speed is x miles/hour, our best estimate of the corresponding engine efficiency level will be $27 - .22x$ miles/gallon. In Table 5.3, we use the equation of the regression line to predict the efficiency level for the speeds 25, 55, and 80. We see that at 25 miles/hour, we predict an efficiency of 21.5 miles/gallon, at 55 miles/hour, we predict 14.9 miles/gallon, and at 80 miles/hour, we predict 9.4 miles/gallon. For any desired[1] value x of speed, a predicted value y of engine efficiency can be worked out in the same way, namely, by using the regression equation $y = 27 - .22x$.

EXERCISES 5.B

1 The manager of an independent supermarket would like to know the relationship, if there is one, between the amount of display space occupied by a local brand of tuna and the dollar value of weekly sales of that item. The amount of space occupied was varied over a period of 6 weeks, and the following data were obtained:

Week Beginning	Display Space, square feet	Total Sales, hundreds of dollars
June 1	5	7
June 8	2	5
June 15	5	7
June 22	8	8
June 29	2	4
July 6	8	9

(a) Draw a scattergram of the data.
(b) Find the equation of the straight line which best fits the data.
(c) Superimpose the graph of the regression line upon the scattergram.
(d) Predict weekly sales for display space of 6 square feet.
2 An agricultural research organization tested a particular chemical fertilizer to try to find out whether an increase in the amount of fertilizer used would

---

[1]The predictions will be more accurate for x values lying within the range of the x data set than for x values lying outside that range.

lead to a corresponding increase in the food supply. They obtained the following data based on seven plots of arable land:

Fertilizer, pounds	Beans, bushels
2	4
1	3
3	4
2	3
4	6
5	5
3	5

(a) Draw a scattergram of the data.
(b) Find the equation of the regression line that would be used to predict the number of bushels of beans obtained from a number of pounds of fertilizer.
(c) Superimpose the graph of the regression line upon the scattergram.
(d) Predict the size of the harvest if 6 pounds of fertilizer are used.
(e) Predict the size of the harvest if no fertilizer is used.

3 One psychologist suspects that there is a connection between the rate of inflation in the economy and the rate of divorce in the general population. In an attempt to find a way to predict the divorce rate from the inflation rate, she collects the following data from records of the past several years:

Inflation rate, %	2	4	5	7	10	12
Divorce rate per 1000 of population	3	7	10	15	25	30

(a) Draw a scattergram of the data.
(b) Find the equation of the regression line.
(c) Superimpose the graph of the regression line upon the scattergram.
(d) If next year's inflation rate looks like it will be 8%, predict next year's divorce rate.

4 A demographer presented the following data to support his theory that high protein diets tend to reduce fertility levels:

Country	Taiwan	Japan	Italy	West Germany	U.S.	Sweden
Protein in diet, grams/day	5	10	15	40	60	70
Birth rate per 100 of population	4	3	3	2	2	1

(a) Draw a scattergram of the data.
(b) Find the equation of the regression line.
(c) Superimpose the graph of the regression line upon the scattergram.
(d) Predict the birth rate in Greece where the typical protein content of the daily diet is 18 grams.

STRENGTH OF THE LINEAR RELATIONSHIP

By use of the formulas for *a* and *b*, a regression line can be found for any set of paired data. Unfortunately, therefore, we can find a regression line even for a set of data that is fundamentally nonlinear, that is to say, a set of data which really does not follow a straight line. Two examples of nonlinear data are presented in Table 5.4, and their scattergrams are illustrated in Fig. 5.9. It is often difficult, especially with large amounts of data, to decide from either the table of data or the scattergram whether the data is truly linear or not. Furthermore, even if the underlying relationship really is a linear one, it would be unreasonable to expect all the data points to fall exactly on the straight line.

To deal with this problem, what is needed is a way of measuring the extent to which a set of data can be considered linear. In particular, we need a way of measuring the strength of the relationship between the data and the regression line calculated from the data. We would use such a measure to answer the question, "Does the regression line present a valid pictorial representation of the behavior of the data?"

Let's look again at a typical set of *n* data pairs $(x_1, y_1), (x_2, y_2), \ldots, (x_n, y_n)$. In general *y* varies; that is to say, the *y* values, $y_1, y_2, \ldots, y_n$, are not all the same. If all the numbers $y_1, y_2, \ldots, y_n$ were the same, they would each be equal to the mean of the group $\bar{y}$. Therefore we can consider the sum of squared deviations from the mean

$$TV = \Sigma(y - \bar{y})^2$$

as a measure of the total variation of *y*. For example, in Table 5.5, we analyze the variation inherent in the engine efficiency data of Table 5.1. From the fourth column of Table 5.5, we can see that the total variation of engine efficiency is

$$TV = \Sigma(y - \bar{y})^2 = 50$$

TABLE 5.4
Two nonlinear sets of data

(a) Parabolic		(b) Logarithmic	
x	y	x	y
8	9	1	1
1	16	3	4
6	1	8	8
3	4	14	9.5
5	0	20	10
2	10	10	9
		5	6
		2	3

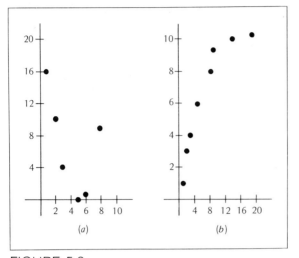

FIGURE 5.9
Scattergrams of nonlinear data.

The sum of squared deviations, SSD, mentioned earlier in this chapter, is actually the total variation of the y values away from the regression line, because it is the sum of the squared deviations of the actual y values from their values $a + bx$ predicted by the regression line. In the engine efficiency example,

$$SSD = \Sigma(y - a - bx)^2 = 1.60$$

as can be seen from the sixth column of Table 5.5.
  Well, the total variation in engine efficiency, as recorded in the data, is 50, while the variation away from the regression line is 1.60. The difference $50 - 1.60 = 48.40$ indicates that there is still substantial variation remaining as part of the total, variation that is not part of the variation away from the regression line. To what can we attribute this remaining variation?

TABLE 5.5
Analysis of the variation of engine efficiency
(Regression line: $y = 27 - .22x$)

Actual Data Points		Total Variation		Variation from Line		Variation with Line
x	y	$\bar{y}$	$(y - \bar{y})^2$	$a + bx$	$(y - a - bx)^2$	$(a + bx - \bar{y})^2$
30	20	16	16	20.4	0.16	19.36
40	18	16	4	18.2	0.04	4.84
50	17	16	1	16.0	1.00	0.00
60	14	16	4	13.8	0.04	4.84
70	11	16	25	11.6	0.36	19.36
Sums	80		50		1.60	48.40

Let's look at the problem from another point of view. Suppose there were no variation at all in y. Then every y value would equal $\bar{y}$, and we would have the total variation $TV = \Sigma(y - \bar{y})^2 = 0$. In fact, however, there is variation in y. Suppose the *only* variation in y were due to the influence of the regression line $y = a + bx$. Then every y would equal its corresponding $a + bx$. The resulting total variation would then be

$$TV = \Sigma(y - \bar{y})^2 = \Sigma(a + bx - \bar{y})^2$$

since y and $a + bx$ would always be the same. The quantity $\Sigma(a + bx - \bar{y})^2$ is therefore the variation in y that can be attributed to the effect of the regression line. We denote this term as

$$VR = \Sigma(a + bx - \bar{y})^2$$

where VR means "variation due to regression." In the far right column of Table 5.5, we have the calculation that

$$VR = \Sigma(a + bx - \bar{y})^2 = 48.40$$

Therefore, we have that

$$TV = SSD + VR$$

because $50 = 1.60 + 48.40$. In words, we can express these assertions as follows:

Total variation of y
= variation away from the regression line
+ variation due to the influence of regression

In our example, then, SSD and VR together account for the entire variation of y, TV. Because $1.60 + 48.40 = 50$, there is no variation left to be attributed to anything else. We might ask the question, "Does this happen in every possible example, or do we have a special case here?" The answer is that TV *always* equals $SSD + VR$. It is an algebraic fact that

$$\Sigma(y - \bar{y})^2 = \Sigma(y - a - bx)^2 + \Sigma(a + bx - \bar{y})^2$$

for *all* sets of paired data.

Because TV is the total variation in y, and VR is the variation in y that can be attributed to the influence of the regression line, the ratio

$$\frac{VR}{TV} = \text{proportion of variation in } y \text{ that can be attributed to influence of regression line}$$

VR/TV is often referred to as "the proportion of the variation in *y* explained by regression," or, simply, the "explained variation." Here

$$\frac{VR}{TV} = \frac{48.40}{50} = .968 = 96.8\%$$

so that 96.8% of the variation in engine efficiency can be explained on the basis of its linear relationship $y = 27 - .22x$ with speed.

The quantity *VR/TV* is called the "coefficient of linear determination" because it measures the extent to which variation in *y* is determined by its linear relationship with *x*. When we say "determined" here, we mean determined algebraically and numerically only; we have not proved, nor can we prove conclusively using mathematics alone, the existence of any type of cause-and-effect relationship between *x* and *y*.

The coefficient of determination is the measure that we have been seeking in this section—a measure of the strength of the linear relationship $y = a + bx$.

As it turns out, it is possible to calculate the coefficient of determination directly from a table like Table 5.2 by means of a relatively straightforward formula. First we calculate

$$\bigstar \quad r = \frac{n\Sigma xy - (\Sigma x)(\Sigma y)}{\sqrt{n\Sigma x^2 - (\Sigma x)^2}\sqrt{n\Sigma y^2 - (\Sigma y)^2}} \tag{5.3}$$

which is called the "coefficient of linear correlation." (In the next section, we will discuss additional uses of the correlation coefficient.) Then the coefficient of linear determination is given by the formula

$$\frac{VR}{TV} = r^2$$

We will show how to use r^2 as the coefficient of determination. In our continuing engine efficiency example, Table 5.2 contains all the basic information needed to compute the correlation coefficient *r*. The calculation is made as follows:

$$r = \frac{n\Sigma xy - (\Sigma x)(\Sigma y)}{\sqrt{n\Sigma x^2 - (\Sigma x)^2}\sqrt{n\Sigma y^2 - (\Sigma y)^2}}$$

$$= \frac{5(3780) - (250)(80)}{\sqrt{5(13,500) - (250)^2}\sqrt{5(1330) - (80)^2}}$$

$$= \frac{18,900 - 20,000}{\sqrt{67,500 - 62,500}\sqrt{6650 - 6400}} = \frac{-1100}{\sqrt{5000}\sqrt{250}}$$

$$= \frac{-1100}{(70.71)(15.81)} = -.984$$

Because $r = -.984$, the coefficient of determination is given by

$$\frac{VR}{TV} = r^2 = (-.984)^2 = .968 = 96.8\%$$

You should observe that we of course get exactly the same value for VR/TV using the correlation coefficient as we obtained earlier by the direct method of Table 5.5.

Before going on to the next topic, let's calculate the coefficients of determination for the two sets of nonlinear data appearing in Table 5.4 and Fig. 5.9. As we will see, they contrast markedly with the 96.8% of the engine efficiency sample.

PARABOLIC DATA

For the parabolic data of Table 5.4, part (a), we first make the preliminary calculations appearing in Table 5.6. To compute the coefficient of determination, i.e., the proportion of variation in y that can be attributed to the linear relationship with x based on the regression line, we then calculate r. We have

$$r = \frac{n\Sigma xy - (\Sigma x)(\Sigma y)}{\sqrt{n\Sigma x^2 - (\Sigma x)^2}\sqrt{n\Sigma y^2 - (\Sigma y)^2}}$$

$$= \frac{6(126) - (25)(40)}{\sqrt{6(139) - (25)^2}\sqrt{6(454) - (40)^2}}$$

$$= \frac{-244}{\sqrt{209}\sqrt{1124}} = \frac{-244}{(14.46)(33.53)} = -.503$$

Therefore $r^2 = (-.503)^2 = .253 = 25.3\%$ of the variation in y can be attributed to the relationship with x given by the regression line. It follows that the regression line fails to account for almost 75% of the variation in y. These numbers provide a strong hint that x and y are really not linearly related. Therefore no attempt

TABLE 5.6
Preliminary calculations for parabolic data

x	y	x^2	y^2	xy
8	9	64	81	72
1	16	1	256	16
6	1	36	1	6
3	4	9	16	12
5	0	25	0	0
2	10	4	100	20
25	40	139	454	126
$n = 6$				

should be made to find the best-fitting straight line, for even that one will not fit very well.

LOGARITHMIC DATA

For the logarithmic data of Table 5.4, part (b), the preliminary calculations appear in Table 5.7. We have

$$r = \frac{n\Sigma xy - (\Sigma x)(\Sigma y)}{\sqrt{n\Sigma x^2 - (\Sigma x)^2}\sqrt{n\Sigma y^2 - (\Sigma y)^2}}$$

$$= \frac{8(536) - (63)(50.5)}{\sqrt{8(799) - (63)^2}\sqrt{8(397.25) - (50.5)^2}}$$

$$= \frac{1106.5}{\sqrt{2423}\sqrt{627.75}} = \frac{1106.5}{(49.22)(25.05)} = .897$$

Therefore the coefficient of determination is

$$r^2 = (.897)^2 = .805 = 80.5\%$$

indicating that the regression line accounts for all but about 20% of the variation in y. It should be observed that the logarithmic data is considerably more linear than the parabolic data having $r^2 = 25.3\%$, but appreciably less linear than the engine efficiency data where $r^2 = 96.8\%$.

One more remark: In addition to its use in simplifying the calculation of the coefficient of determination, the correlation coefficient r immediately gives information about the direction of the trend of the data. In particular, if r turns out to be negative, this means that the regression line slopes downward to the right, in the manner of the scattergram in Fig. 5.8. If, on the other hand, r is positive, the line slopes upward to the right, as in Figs. 5.3, 5.6, and 5.9b. The reason for this association of r with slope is the close mathematical relationship between

TABLE 5.7
Preliminary calculations for logarithmic data

x	y	x^2	y^2	xy
1	1	1	1	1
3	4	9	16	12
8	8	64	64	64
14	9.5	196	90.25	133
20	10	400	100	200
10	9	100	81	90
5	6	25	36	30
2	3	4	9	6
63	50.5	799	397.25	536
n = 8				

the formulas for r and b. In fact

$$r = \frac{n\Sigma xy - (\Sigma x)(\Sigma y)}{\sqrt{n\Sigma x^2 - (\Sigma x)^2}\sqrt{n\Sigma y^2 - (\Sigma y)^2}} = \frac{b\sqrt{n\Sigma x^2 - (\Sigma x)^2}}{\sqrt{n\Sigma y^2 - (\Sigma y)^2}}$$

so that the algebraic sign of r ($+$ or $-$) is always the same as the sign of b. But, for $y = a + bx$, when b (and so r) is negative, y grows smaller as x grows larger, and so the line slopes downward to the right. Analogously, if b (and so r) is positive, y grows larger as x grows larger, and so the line slopes upward to the right. Therefore, taking note of the sign of r and its square, the coefficient of determination, we can form a good mental picture of the regression line and how well the data fit it.

EXERCISES 5.C

1   Based on the data presented in Exercise 5.B.1, what proportion of the variation in total sales can be explained by a linear relationship between total sales and display space?
2   From the data of Exercise 5.B.2, how much of the variation in yield of beans can be attributed to a linear relationship with the amount of fertilizer?
3   On the basis of the data given in Exercise 5.B.3 and the linear relationship described there, how much of the variation in the divorce rate can be ascribed to corresponding variations in the inflation rate?
4   Using the demography data of Exercise 5.B.4, what percentage of the variation in birth rates can be explained in terms of variations in the protein content of national diets and its linear relationship with the national birth rate?
5   Coronado Nautotronics of San Diego, bidding for the contract to produce the radar displays for the Navy's new fleet of patrol hydrofoils, collects the following information in an attempt to determine the cost curve for the radar displays:

Quantity produced	10	50	100	160	200	320	630	800
Total cost of production run	7.0	8.5	9.0	9.4	9.5	10.0	10.5	10.8

(a)   Draw a scattergram of the data.
(b)   Calculate the coefficient of linear determination.

6   An economist wants to determine the daily demand equation for rolled steel in a small industrial town. She collects the following data, relating the price with the quantity of rolled steel that can be sold at that price:

Tons of Rolled Steel that Can Be Sold	Price per Ton, hundreds of dollars
1.0	50.10
2.0	12.60
2.5	8.00
4.0	3.20
5.0	2.00
6.3	1.25

(a) Draw a scattergram of the data.

(b) Calculate the coefficient of linear determination.

7 In an attempt to develop an aptitude test measuring an individual's aptitude for pursuing a career in computer programming, a psychologist first compares mathematical aptitude scores of already employed computer programmers with their job performance ratings. The data follow:

Person	A	B	C	D	E	F	G
Math aptitude score	2	5	0	4	3	1	6
Job performance rating	8	5	1	8	9	5	1

(a) Draw a scattergram of the data.

(b) Calculate the coefficient of linear determination, and find out what proportion of variation in job performance ratings can be explained on the basis of a linear relationship with mathematical aptitude scores.

SECTION 5.D
A TEST FOR LINEARITY

The correlation coefficient (for historical reasons, sometimes referred to as the "Pearson product-moment correlation coefficient") can also be used to test a statistical hypothesis. The "correlation test" decides whether or not the population from which a sample of data points was drawn can reasonably be considered, at an appropriate significance level, to be linear.

As we have noted in earlier situations of statistical inference, it is necessary for the validity of a small-sample correlation test of linearity that both the x and y values have been drawn from populations which can reasonably be assumed to have normal distributions. Under the required assumption of normality, then, we have an underlying true coefficient of linear correlation ρ (pronounced "row") of which r is merely a sample estimate based on n data points. We want to test the hypothesis

H: $\rho = 0$ (no correlation—data is nonlinear)

against the alternative

A: $\rho \neq 0$ (significant correlation—data exhibits significant linear trend)

If ρ were near 0, r would most likely be small, and so $VR/TV = r^2$ would represent a very small percentage of the total variation of y. This would indicate that the relationship between x and y, if one exists, is probably nonlinear. On the other hand, a significantly large value of ρ, in either the positive or negative direction, would be consistent with a high percentage of the variation of y's being accounted for by the regression line, and so rejection of **H:** $\rho = 0$ in favor of

TABLE 5.8
Correlation coefficients of sets of data

Data Set	n	r
Engine efficiency	5	−.984
Parabolic	6	−.503
Logarithmic	8	.897

A: $\rho \neq 0$ would tend to indicate the existence of a significant linear relationship in the underlying population. For simplicity, we can abbreviate our testing problem as

H: $\rho = 0$ (nonlinear)

versus

A: $\rho \neq 0$ (linear)

To test **H** against **A**, we use the test statistic

$$\bigstar \quad t = \frac{r\sqrt{n-2}}{\sqrt{1-r^2}} \tag{5.4}$$

which has the t distribution with $n - 2$ degrees of freedom. We would therefore reject **H** (nonlinearity) in favor of **A** (linearity) at significance level α if

$$|t| > t_{\alpha/2}[n-2]$$

Let's apply the technique to each of the three sets of data discussed in the previous section. Suppose we consider various values of α simultaneously, so that we may compare the sets of data with regard to their linearity. From the previous section, let's recall the information in Table 5.8. (It is important to observe that r will always fall between -1 and $+1$ for any set of data. This is because r^2 is a proportion and so must always lie between 0 and 1. A value of r outside the range -1 to $+1$ signals the presence of a computational error.) The next step in carrying out the test of linearity is to look up the comparison values $t_{\alpha/2}[n-2]$. We obtain these from Table A.4 in the back of the book for three levels of α and compile them in Table 5.9.

TABLE 5.9
Comparison t values for correlation test
$t_{\alpha/2}[n-2]$

n	n − 2	$\alpha = .10$	$\alpha = .05$	$\alpha = .01$
5	3	2.353	3.182	5.841
6	4	2.132	2.776	4.604
8	6	1.943	2.447	3.707

It remains only to calculate, from Table 5.8, the numerical value of the test statistic in Eq. (5.4) and to compare it against the corresponding t value listed in Table 5.9.

ENGINE EFFICIENCY DATA

For the engine efficiency example, we have

$$t = \frac{r\sqrt{n-2}}{\sqrt{1-r^2}} = \frac{-.984\sqrt{5-2}}{\sqrt{1-(-.984)^2}} = \frac{-.984\sqrt{3}}{\sqrt{1-.968}} = \frac{-.984\sqrt{3}}{\sqrt{.0317}}$$

$$= \frac{(-.984)(1.732)}{.178} = -9.6$$

We are to reject

H: $\rho = 0$ (nonlinear)

in favor of

A: $\rho \neq 0$ (linear)

if

$$|t| > t_{\alpha/2}[n-2]$$

Clearly $|t| = 9.6$ exceeds $t_{\alpha/2}[3]$ for all reasonable levels of α, as can be seen from the top row of numbers in Table 5.9. This means that whatever your selected level of α, within a reasonable range, you would be led to reject **H** in favor of **A**, concluding that in the engine efficiency problem, efficiency is related to speed by a linear relationship. (That relationship is, of course, the one given by the regression line $y = 27 - .22x$ of Sec. 5.B.)

PARABOLIC DATA

For the parabolic data, Table 5.8 indicates that

$$t = \frac{r\sqrt{n-2}}{\sqrt{1-r^2}} = \frac{-.503\sqrt{6-2}}{\sqrt{1-(-.503)^2}} = \frac{-.503\sqrt{4}}{\sqrt{1-.253}} = \frac{(-.503)(2)}{\sqrt{.747}}$$

$$= \frac{-1.006}{.864} = -1.164$$

Here $|t|$ 1.164, which fails to exceed $t_{\alpha/2}[4]$ recorded in Table 5.9 for any reasonable level of α. Therefore, we would fail to reject **H** at every reasonable α, concluding that we have no evidence to indicate the presence of a linear relationship in the underlying population.

LOGARITHMIC DATA

For the logarithmic data, we see from Table 5.8 that

$$t = \frac{r\sqrt{n-2}}{\sqrt{1-r^2}} = \frac{.897\sqrt{8-2}}{\sqrt{1-(.897)^2}} = 4.971$$

Here $|t| = 4.971$ exceeds the values of $t_{\alpha/2}[6]$ recorded in Table 5.9 for every reasonable α. Therefore we would reject **H** and conclude, at every reasonable level α, that there is significant evidence of a linear relationship between x and y in the underlying population. Recalling that when $\alpha = .01$, there is a 1% chance of rejecting **H** when it's really true, the fact that $|t| = 4.971$ for the logarithmic data of Fig. 5.9b means that if the underlying population were strongly nonlinear, there would be less than a 1% chance of having $|t|$ as high as 4.971. Because $|t|$ does turn out as high as 4.971, we know that the logarithmic data cannot be considered as strongly nonlinear. In fact, looking at its scattergram in Fig. 5.9b, we see that it is only slightly nonlinear. While a logarithmic curve would describe the data best, our correlation test of linearity shows that the regression line would not be very bad.

If it is known that the x and y data points come from underlying populations that are normally distributed, the correlation test of linearity should be performed before any calculation of the regression line is made. Such a procedure will protect against the waste of time and effort involved in calculating a regression line that will be essentially useless in case the data is later judged to be nonlinear. When it is not reasonable to assume normality in the underlying populations, the best we can do is to use the coefficient of linear determination. Of course, in the latter situation, we cannot obtain the degree of precision in significance levels that is available in the case of normality.

EXERCISES 5.D

1 At level of significance $\alpha = .05$, can we consider the relationship between display space and total sales as discussed in Exercises 5.B.1 and 5.C.1 to be a linear one?
2 At significance level $\alpha = .05$, is it reasonable to assume that the relationship between amount of fertilizer and bushels of beans studied earlier in Exercises 5.B.2 and 5.C.2 is linear?
3 At level $\alpha = .01$, does it make sense to use a linear equation to describe the

relationship between inflation rate and divorce rate based on the data analyzed in Exercises 5.B.3 and 5.C.3?

4   Using a level of significance of $\alpha = .01$, does the relationship between protein content of diet and national birth rate based on the data of Exercises 5.B.4 and 5.C.4 appear to be linear?

5   At level $\alpha = .05$, does the cost curve based on the data of Exercise 5.C.5 seem to be linear?

6   Can the demand curve of Exercise 5.C.6 be considered linear at level $\alpha = .05$?

7   At level of significance $\alpha = .10$, can we validly use a linear equation to describe the relationship between mathematical aptitude and job performance in computer programming based on the data of Exercise 5.C.7?

SECTION 5.E
A COMPLETELY WORKED-OUT EXAMPLE

The many new topics introduced in this chapter often appear in practice as merely the components of a single problem and its comprehensive solution. This essential unity among the various techniques has perhaps been hidden from you, since the techniques have been scattered over several pages of text. In the final section of this chapter, we present an example of a regression-correlation problem, completely solved in a logical sequence of steps from beginning to end.

Example 5.1 Bacterial Growth   A biologist conducts an experiment aimed at determining how the growth rate of a certain type of bacteria increases with the passage of time. She sets up six cultures of the bacteria, studies each one's growth for a different time period, and records their growth rates at the conclusion of the respective time periods. The data follow:

Culture	A	B	C	D	E	F
Time period, hours	3	5	1	7	5	9
Growth rate, units/hour	10	13	8	13	11	15

The following questions are of interest to the biologist:

1   What proportion of the variation in growth rates among the bacteria can be attributed to a linear relationship with corresponding variations in the time periods?

2   What does the scattergram of the data look like?

3   At significance level $\alpha = .05$, can we consider the growth rate to be linearly related to the length of the time period?

4   If so, what is the equation of the regression line which will predict growth rates for various given time periods?

5 Predict the growth rate of a culture of the bacteria after a time period of 12 hours.

SOLUTION

1 To find the proportion of variation in growth rates that can be explained on the basis of variations in the time periods, we have to compute the coefficient of linear determination. We first set up the table of preliminary calculations, Table 5.10.

Here the growth rate, the variable to be predicted, is denoted by y, while the time period is denoted by x.

The coefficient of determination is r^2, where

$$r = \frac{n\Sigma xy - (\Sigma x)(\Sigma y)}{\sqrt{n\Sigma x^2 - (\Sigma x)^2}\sqrt{n\Sigma y - (\Sigma y)^2}}$$

$$= \frac{6(384) - (30)(70)}{\sqrt{6(190) - (30)^2}\sqrt{6(848) - (70)^2}}$$

$$= \frac{2304 - 2100}{\sqrt{1140 - 900}\sqrt{5088 - 4900}} = \frac{204}{\sqrt{240}\sqrt{188}}$$

$$= \frac{204}{(15.5)(13.7)} = .96$$

As $r^2 = (.96)^2 = .922$, 92.2% of the variation in growth rate can be attributed to whatever linear relationship may exist between growth rate and time period.

2 The scattergram of the growth rate data appears in Fig. 5.10.

3 We test the hypothesis

H: $\rho = 0$ (nonlinear)

TABLE 5.10
Preliminary calculations for growth rate example

Time Period x, hours	Growth Rate y, units/hour	x^2	y^2	xy
3	10	9	100	30
5	13	25	169	65
1	8	1	64	8
7	13	49	169	91
5	11	25	121	55
9	15	81	225	135
Sums 30 n = 6	70	190	848	384

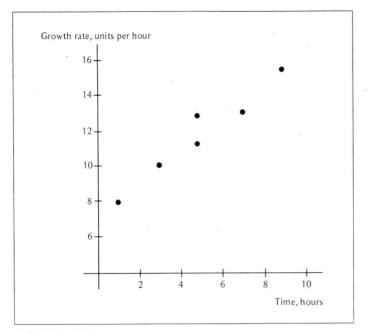

FIGURE 5.10
Scattergram of growth rate data.

versus

A: $\rho \neq 0$ (linear)

at level $\alpha = .05$ by calculating

$$t = \frac{r\sqrt{n - 2}}{\sqrt{1 - r^2}}$$

and rejecting **H** if

$$|t| > t_{\alpha/2}[n - 2] = t_{.025}[4] = 2.776$$

We have

$$t = \frac{r\sqrt{n - 2}}{\sqrt{1 - r^2}} = \frac{.96\sqrt{6 - 2}}{\sqrt{1 - (.96)^2}} = \frac{.96\sqrt{4}}{\sqrt{1 - .9216}} = \frac{1.92}{\sqrt{.0784}}$$

$$= \frac{1.92}{.28} = 6.857$$

Therefore $|t| = 6.857 > 2.776$, and so we reject **H** and conclude at level $\alpha = .05$ that growth rate and length of time period seem to be linearly related.

4 The regression line has equation $y = a + bx$, where y represents growth rate (the variable to be predicted), and x represents length of time period. Here

$$b = \frac{n\Sigma xy - (\Sigma x)(\Sigma y)}{n\Sigma x^2 - (\Sigma x)^2} = \frac{204}{240} = .85$$

where we have obtained the quantities $n\Sigma xy - (\Sigma x)(\Sigma y) = 204$ and $n\Sigma x^2 - (\Sigma x)^2 = 240$ not by computation, but by merely looking their values up among the calculations we made to find the correlation coefficient r. Notice that the numerator in the formula for b is exactly the same as the numerator in the formula for r, while the denominator of b is one of the factors appearing under a square root sign in the expression for r. (In view of the relationship between b and r, therefore, we are able to simplify the calculation of b somewhat.) Then

$$a = \frac{\Sigma y - b\Sigma x}{n} = \frac{70 - (.85)(30)}{6} = \frac{70 - 25.5}{6}$$

$$= \frac{44.5}{6} = 7.42$$

The regression line therefore has equation $y = 7.42 + .85x$. For illustrative purposes, as well as to provide a check on our calculations, we often superimpose the graph of the regression line upon the scattergram. To do

FIGURE 5.11
Regression line superimposed on scattergram.

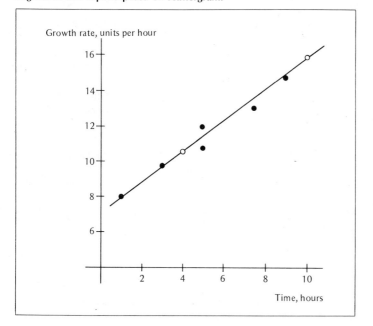

this, we choose two values of x within the range of the original data, say x = 4 and x = 10. If x = 4, then y = 7.42 + (.85)(4) = 10.82, so that (4, 10.82) lies on the regression line, and if x = 10, then y = 7.42 + (.85)(10) = 15.92, so that (10, 15.92) lies on the regression line. The regression line is shown superimposed on the scattergram in Fig. 5.11.

5 To predict the growth rate of a culture of the bacteria after a time period of 12 hours, we set x = 12 in the regression equation y = 7.42 + .85x. We therefore predict that after 12 hours a culture will be growing at the rate of 7.42 + (.85)(12) = 17.62 units per hour.

EXERCISES 5.E

1 To check on the possibility that the slope of a delta region can be accurately predicted from a knowledge of the typical size of the stones found there, a geographer gathered stones from various locations in a delta region of West Baffin Island. The typical size of the stones found at each location was then compared with the cotangent of the angle of slope of the delta there. (A larger cotangent corresponds to a smaller angle of slope.) The resulting data follow:

Typical Size of Stones, diameter in meters	Slope of Delta, 1% of cotangent of angle
.5	3.0
3.0	.5
1.0	2.0
2.5	1.0
2.0	1.5

(a) What proportion of the variation in delta slope can be explained on the basis of a linear relationship with size of stones?
(b) Draw a scattergram of the data.
(c) At a level of significance α = .02, can the relationship between size of stones and slope of delta be considered linear?
(d) Determine the equation of the regression line which predicts delta slope in terms of stone size, and superimpose the graph of the regression line upon the scattergram.
(e) Find the best prediction for 1% of the cotangent of the angle of slope at a location where the typical stone has diameter 1.5 meters.

2 The manager of a salmon cannery suspects that the demand for his product is closely related to the disposable income of his target region. To test out his suspicion, he collects the following data for 1976:

Region	Disposable Income, millions of dollars	Sales Volume, thousands of cases
A	10	1
B	20	3
C	40	4
D	50	5
E	30	2

(a) What percentage of the variation in sales volume among regions can be attributed to a linear relationship with disposable income of the regions?

(b) Draw a scattergram of the data.

(c) At level $\alpha = .05$, do the data indicate the existence of a linear relationship between disposable income and sales volume?

(d) Develop the regression equation which can be used to predict sales volume in a region from a knowledge of the region's disposable income, and superimpose the graph of the regression line upon the scattergram.

(e) Find the best prediction for the sales volume in a region whose disposable income is 25,000,000 dollars.

3 A sociologist employed by HEW is assigned the task of finding the relationship, if there is any, between the level of an individual's formal education and whether or not that person is currently unemployed. One of her methods of investigation centers on a comparison between the various levels of formal education and the unemployment rate among individuals at that educational level. In the data which follow, the years 0 to 8 are grade school, 9 to 12 are high school, 13 to 16 are college, 17 to 18 are master's degree study, and 19 to 24 are doctoral degree study.

Educational Level, years	Unemployment Rate, %
4	30
8	20
12	6
14	10
18	3
24	6

The researcher would like to use the data to develop a method of predicting a person's chances of being unemployed from that person's educational level.

(a) What proportion of the variation in unemployment rate among the various educational levels can be accounted for by a linear relationship with educational level?

(b) Draw a scattergram of the data.

(c) At level $\alpha = .05$, can the sociologist conclude that the relationship between educational level and unemployment rate is a linear one?

(d) Find the equation of the regression line for predicting unemployment rate from educational level, and superimpose the graph of the regression line upon the scattergram.

(e) Predict the unemployment rate among persons having 10 years of formal education.

4 A pharmaceutical researcher has to know, as precisely as possible, the effect that a new drug will have on the human pulse rate. To investigate this effect, he administers different dosages of the drug to each of seven randomly selected patients, and he notes the increase in their pulse rates 1 hour later. The data follow:

Dosage, cubic centimeters	1.5	2.0	2.5	1.5	3.0	2.0	2.5
Change in pulse rate	8	10	14	10	15	11	12

(a) On the basis of a linear relationship, what percentage of the variation in pulse rate change can be attributed to variations in dosage?

(b) Construct a scattergram of the data.

(c) At level $\alpha = .05$, is there sufficient evidence to conclude that the relationship is really linear?

(d) Determine the equation of the regression line capable of predicting pulse rate change from a knowledge of the dosage, and superimpose the graph of the regression line upon the scattergram.

(e) Predict the change in pulse rate corresponding to a dosage of 4.0 cubic centimeters.

5 In a study of the advertising budgets of small businesses, a consultant to the Small Business Administration collects the following data relating the size of a business' advertising budget with that business' total sales volume:

Advertising budget, hundreds of dollars	6	4	10	1	7	5	7
Total sales volume, thousands of dollars	50	60	100	30	60	60	40

(a) On the basis of the given data, what proportion of variation among total sales volumes of businesses can be explained by a linear relationship between the advertising budget and the sales volume?

(b) Draw a scattergram of the data.

(c) At level $\alpha = .10$, can the relationship described by the data be considered linear?

(d) Can the relationship be considered linear at a level of significance of $\alpha = .05$?

(e) Find the equation of the straight line which best predicts sales volume from a knowledge of the advertising budget, and superimpose the graph of the line upon the scattergram.

(f) Predict the total sales volume of a small business that spends 800 dollars on advertising.

(g) Predict the total sales volume of a small business that spends 300 dollars on advertising.

SUMMARY AND DISCUSSION

In developing the techniques introduced in Chap. 5, our goal has been to discover the relationship between two sets of data for the purpose of eventually predicting one of the quantities from a knowledge of the other. The simplest and most fundamental measure of that relationship is, as we have seen, the coefficient of linear determination, which indicates the extent to which a straight line having equation $y = a + bx$ describes the data. A statistic closely tied to the coefficient of linear determination, Pearson's correlation coefficient r, can be used to test hypotheses involving the strength of a linear relationship in cases where the data points come from a normally distributed population. If our indicators seem to show that a linear relationship adequately describes the data, the next step would be to specify the regression line, the line which most closely approximates all the original data points simultaneously. After we calculated the best-fitting (least-squares) line, we showed how to use the line for the purpose of predicting one quantity from the other.

SUPPLEMENTARY EXERCISES

1 As part of an analysis of the relationship between smoking and absenteeism, the following data were obtained, relating an individual's number of packs smoked per day with the individual's number of days absent from his or her job.

Individual	A	B	C	D	E	F	G	H
Number of packs smoked per day	.5	1.5	2.0	.5	.0	1.0	3.5	.0
Number of days absent per year	4	8	15	0	3	10	20	0

(a) What proportion of the variation in absenteeism among individuals can be explained on the basis of a linear relationship between absenteeism and amount of smoking?

(b) Draw a scattergram of the data.

(c) At significance level $\alpha = .05$, is it reasonable to assume, from the data collected, that the relationship between smoking and absenteeism is linear?

(d) Calculate the equation of the regression line for predicting an individual's level of smoking, and superimpose the graph of the regression line upon the scattergram.

(e) Predict the number of days an employee who smokes five packs a day will be absent.

2 A sociological study undertaken to investigate the relationship between the population density of a metropolitan area and that area's crime rate yields the following data:

Metropolitan Area	#1	#2	#3	#4	#5	#6
X Population density, thousands per square mile	20	15	30	5	10	10
Y Crimes reported per 10,000 of population	10	8	12	1	4	5

(a) According to the results of the study, what proportion of variation in the crime rate among various metropolitan areas can be ascribed to variations in population density by way of a linear relationship?

(b) Draw a scattergram of the data.

(c) Decide using a level of significance $\alpha = .05$ whether or not the relationship between population density and crime rate, as described by the data, can reasonably be considered to be linear.

(d) Determine the equation of the regression line which best accounts for the relationship between population density and crime rate, and superimpose its graph upon the scattergram.

(e) Find the best prediction for the crime rate in a metropolitan area having population density of 25,000 per square mile.

3 Some investors think that prices on the New York Stock Exchange are related to prices on the London Gold Exchange. To test out this theory, one financial analyst compared the Dow-Jones Industrial Average with the price of gold in London on five randomly selected days. He came up with the following data in an attempt to find out if he could predict changes in the Dow-Jones Average from the behavior of the London gold prices which close several hours earlier due to time zone differences.

London Gold Prices	Dow-Jones Industrial Average
2.00	560
1.90	600
1.60	840
1.80	700
1.70	840

(a) According to the data, how much of the variation in the Dow-Jones Average is attributable to a linear relationship with the price of gold in London?

(b) Draw a scattergram of the data.

(c) At level $\alpha = .01$, do the data substantiate the existence of a linear relationship between the Dow-Jones Average and the price of gold in London?

(d) Calculate the regression line for predicting the Dow-Jones Average from the price of London gold, and superimpose the graph of the regression line upon the scattergram.

(e) Predict the Dow-Jones Average if the price of gold in London were to rise to 2.20.

4 A meteorologist studying the relationship between altitude and temperature accumulates the following data at six military airfields around the state:

Airfield	A	B	C	D	E	F
Altitude, hundreds of feet above sea level	9	70	7	4	50	20
High temperature, °F	90	50	100	110	70	80

(a) What proportion of the variation in temperature can be explained by a linear relationship with altitude?

(b) Draw a scattergram of the data.

(c) At level $\alpha = .05$, can we assert that the relationship between altitude and high temperature is a linear one?

(d) Determine the equation of the regression line expressing high temperature in terms of altitude, and superimpose its graph upon the scattergram.

(e) Predict the high temperature at an altitude of 3500 feet above sea level.

5 A manufacturer of chemicals for use in the pharmaceutical industry conducts a study of the time it takes after receipt of an order for the chemicals

to reach the customer's plant. Because some of the chemicals deteriorate over time, an accurate estimate of the transit time is needed. The following data compare the rail distance from the manufacturer to the customer with the transit time of the most recent shipment to that customer:

Customer	#1	#2	#3	#4	#5	#6
Rail distance, hundreds of miles	5	4	1	3	2	5
Transit time, days	6.0	5.0	3.0	4.5	4.0	5.5

(a) Find the proportion of variation in transit times that can be attributed to a linear relationship with the rail distances.
(b) Draw a scattergram of the data.
(c) At level $\alpha = .05$, would we be justified in using a linear equation to predict the transit time from a knowledge of the rail distances?
(d) Find the equation of the regression line, and superimpose its graph upon the scattergram.
(e) Customer #7 is located 700 miles from the manufacturer. Predict the time it will take a shipment of chemicals to arrive at his warehouse.

6 A botanist has collected the following data on the height (in centimeters) of a certain plant at several ages (in weeks after germination):

Age	1	2	4	6	7
Height	10	30	30	40	50

(a) What proportion of the variation in height seems to be attributable to the age of the plant by way of a linear relationship between height and age?
(b) Draw a scattergram of the data.
(c) Decide at level $\alpha = .05$ whether or not the relationship between age and height can be considered linear.
(d) Determine the equation of the regression line which expresses height of the plant in terms of its age, and superimpose the graph of the regression line upon the scattergram.
(e) Predict the height of the plant 10 weeks after germination.

7 Some economists believe that as the prime interest rate increases, the value of stocks declines. The following data compare the prime rate with the value of General Nucleonics common stock at five randomly selected times during the past several years:

Prime rate, %	10	6	5	7	12
GN stock prices, dollars per share	60	80	90	70	50

(a) What proportion of the variation in GN stock prices can be accounted for by a linear relationship with the prime interest rate?
(b) Draw a scattergram of the data.
(c) At level $\alpha = .05$, can the relationship between the prime rate and the price of GN stock be considered linear?

(d) Determine the equation of the regression line which best predicts GN stock prices from the prime rate, and superimpose the graph of the regression line upon the scattergram.

(e) Predict the value of GN stock if the prime rate were to rise to 15%.

8 One sociologist believes that an increase in the divorce rate foreshadows an increase in the crime rate. He obtains the following data on the divorce and crime rates per 10,000 population for five metropolitan areas:

Area	Divorce Rate	Crime Rate
A	2	1
B	6	4
C	8	6
D	4	3
E	10	6

(a) What proportion of the variation in the crime rate from area to area can be explained on the basis of a linear relationship between the crime rate and the divorce rate?

(b) Draw a scattergram of the data.

(c) At a level of significance of $\alpha = .05$, is it reasonable to consider crime rate and divorce rate to be linearly related?

(d) Determine the equation of the regression line which best predicts crime rate from divorce rate, and superimpose the graph of the regression line upon the scattergram.

(e) Predict the crime rate in a metropolitan area where the divorce rate is 5 per 10,000 population.

9 A record of maintenance cost is assembled on six identical metal-stamping machines of different ages in an attempt to find out the relationship, if there is any, between the age of a machine (in years) and the monthly maintenance cost (in dollars) required to keep it in peak operating condition. Such information would be useful in deciding when it is most economical to replace the old machines with new ones. Data on six randomly selected machines follow:

Machine Serial No.	927	901	887	923	891	906
Age, years	2	1	3	4	4	6
Maintenance cost, dollars per month	10	10	30	30	40	50

(a) What proportion of the variation in maintenance cost can be attributed, by the way of a linear relationship, to variations in age?

(b) Draw a scattergram of the data.

(c) At level $\alpha = .05$, can the relationship between maintenance cost and machine age be considered linear?

(d) Find the regression equation which predicts monthly maintenance cost in terms of machine age, and superimpose the graph of the regression line upon the scattergram.

(e) Predict the monthly maintenance cost of a 7-year-old machine.

10 The following table gives the wholesale price index and the index of industrial production in the United States on July 1 of each year of a 5-year
period. Knowledge of the relationship between the two indices would be
helpful to an economist attempting to forecast business conditions.

Year	Wholesale Prices	Industrial Production
#1	80	110
#2	80	90
#3	90	80
#4	80	100
#5	70	120

(a) What proportion of variation in the index of industrial production
seems to be explained by a possible linear relationship with the
wholesale price index?
(b) Draw a scattergram of the data.
(c) At level $\alpha = .10$, decide whether or not it is reasonable to consider the
two indices to be linearly related.
(d) Find the equation of the regression line which can be used to predict
the index of industrial production from a knowledge of the wholesale
price index, and superimpose the graph of the regression line upon the
scattergram.
(e) Predict the index of industrial production when the wholesale price
index stands at 100.

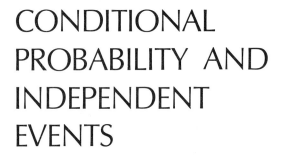

CONDITIONAL PROBABILITY AND INDEPENDENT EVENTS

In earlier sections of the text, we have often referred to the probability of various occurrences. Up to now, however, we have used probability notions and concepts primarily as background for statistical analyses such as confidence intervals and hypothesis testing. Our point of view in this chapter will be different. Here we will be studying directly the probabilities of various events, and we will be calculating the probabilities of related events based on some collected information or data. By "probability of an event," we usually mean the proportion of times the event can be expected to occur over a long period of time. Sometimes, however, events cannot be repeated (for example, in attempting to find the probability that an earthquake will destroy Los Angeles), and then the probability will have to mean our view of the chances of the occurrence of the event, based on prior information in our possession.

SECTION 6.A
EVENTS

The collection of all possible outcomes of an experiment, survey, or other method of data collection is called the "sample space" of the experiment. To simplify the explanation of the concepts involved, we will illustrate all the basic concepts of probability by use of a "die," sometimes called a probability cube, a picture of which can be found in Fig. 6.1. (The plural of die is dice.) We will make a detailed analysis of the following experiment for the purpose of introducing the concepts of probability: We roll the die once only, and we note the number of dots on the side facing upward when the die comes to a stop. If the die comes to a stop as in Fig. 6.1, with the one-dot side facing upward, we say we have rolled a 1. The sample space of this experiment, the collection of all possible outcomes, consists of the numbers 1, 2, 3, 4, 5, and 6. We denote this set as follows:

$$S = \{1, 2, 3, 4, 5, 6\}$$

By an "event," we mean a collection of some, but not necessarily all, of the possible outcomes of an experiment. For example, in the probability cube experiment, we can consider the event that we roll an even number. This event can be symbolized by

$$E = \{2, 4, 6\}$$

Other events that may be of interest are the events

$L = \{1, 2\}$ = we roll a very low number

$H = \{5, 6\}$ = we roll a very high number

$A = \{1, 2, 3, 4\}$ = we roll a not-too-high number

FIGURE 6.1
Die (probability cube).

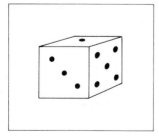

There is an "arithmetic of events," a way of combining events, that is somewhat analogous to addition and multiplication of numbers. Those of you having a background in set theory or the new mathematics might recognize some of the concepts of the arithmetic of events.

1 The "intersection of two events" is the set consisting of all outcomes which are simultaneously in both the events. For example, the intersection of events E and L above, abbreviated $E \cap L$, where the symbol $\cap$ is pronounced "intersect," is

$$E \cap L = \{2\}$$

because 2 is the only outcome found in both E and L.

2 The "union of two events" is the set consisting of all outcomes which are either in one or both the events. For example, the union of events E and L, abbreviated $E \cup L$, where the symbol $\cup$ is pronounced "union," is

$$E \cup L = \{1, 2, 4, 6\}$$

because 1 is in L, 4 and 6 are in E, and 2 is in both L and E.

3 The "complement of an event" is the set consisting of all those outcomes in the sample space S which are not in the event itself. For example, the complement of the event A above, abbreviated A^c, which is pronounced "A complement," is

$$A^c = \{5, 6\}$$

because $A = \{1, 2, 3, 4\}$, and $S = \{1, 2, 3, 4, 5, 6\}$, so that 5 and 6 are the outcomes in S which are not in A. In view of the fact that we have $H = \{5, 6\}$ above, you should notice that $A^c = H$, because A^c and H consist of exactly the same outcomes.

The above three operations together constitute the basic structure of the arithmetic of events. One more item is also used extensively in this arithmetic, namely, the empty event. The empty event, denoted by the symbol ϕ (the Greek letter, lowercase "phi," pronounced "fee" or "fie"), is the event which consists of no outcomes at all. It plays a role in the arithmetic of events analogous to the role played by the number 0 in ordinary arithmetic of numbers. We note that

1 $L \cap H = \phi$ because there are no outcomes which are both in L and in H, as $L = \{1, 2\}$ and $H = \{5, 6\}$.

2 $A \cap A^c = \phi$ because there are no outcomes which are in both A and A^c, as A^c consists of exactly those outcomes which are not in A.

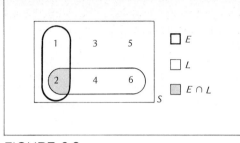

FIGURE 6.2
Intersection of events.

The two events L and H are said to be "mutually exclusive" or "disjoint" because $L \cap H = \phi$. In fact, if C and D are any two events such that $C \cap D = \phi$, then they are said to be disjoint events.

By means of pictures called "Venn diagrams," we can illustrate the concepts of intersection, union, complement, and disjoint events. We do this in Figs. 6.2, 6.3, 6.4, and 6.5.

There is an important relationship among events which we will need to use when we get ready to solve problems in applied fields of work. This relationship is an equation expressed in the language of the arithmetic of events and involves the three concepts of intersection, union, and complement. Suppose we have two events, labeled E and F, as in the Venn diagram of Fig. 6.6. Then

$E \cap F$ = set of outcomes in E which are also in F

$E \cap F^c$ = set of outcomes in E which are not in F

Now, these two sets together make up E, since an outcome in E must be either in F or not in F. (In fact, every outcome in the sample space is either in F or not in

FIGURE 6.3
Union of events.

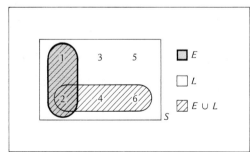

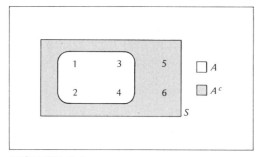

FIGURE 6.4
Complement of an event.

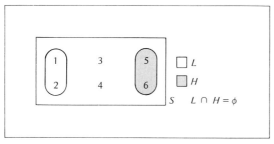

FIGURE 6.5
Disjoint events.

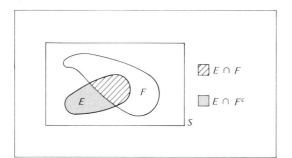

FIGURE 6.6
$E = (E \cap F) \cup (E \cap F^c)$.

F.) Therefore, as Fig. 6.6 illustrates, E is the union of the two sets $E \cap F$ and $E \cap F^c$. We can express this in probability symbolism as follows:

★ $E = (E \cap F) \cup (E \cap F^c)$ (6.1)

When we begin to use probability methods to solve problems in the various applied fields of interest, we shall need to use the above equation.

1 Consider the experiment of tossing a pair of dice, one of them red and the other green. The outcome of the experiment is considered to be the total number of dots facing upward on both dice together.
(a) Explain why the sample space of this experiment is the set $S = \{2, 3, 4, 5, 6, 7, 8, 9, 10, 11, 12\}$.
(b) Draw a Venn diagram of the sample space and the following events:

$$W = \{7, 11\} \qquad\qquad E = \{2, 4, 6, 8, 10, 12\}$$

$$L = \{2, 3, 12\} \qquad\qquad D = \{3, 5, 7, 9, 11\}$$

(c) List the outcomes of the following events:

$W \cup L$	$W^c \cap L^c$
$W \cap L$	E^c
W^c	$E^c \cap D$
L^c	$E^c \cup D$
$W^c \cup L^c$	

2 (a) In the situation of Exercise 6.A.1, list the outcomes of each of the events $(L \cap E)^c$ and $L^c \cup E^c$.
(b) Construct a Venn diagram which illustrates the fact that, for any two events A and B, it is always true that $(A \cap B)^c = A^c \cup B^c$.

3 Consider the experiment of rolling three dice simultaneously, one red, the second green, and the third white. The outcome of the experiment is considered to be the total number of dots facing upward on all three dice together.
(a) List the sample space of this experiment.
(b) Draw a Venn diagram of the sample space and each of the following events:

$$W = \{7, 11, 17\}$$

$$L = \{3, 4, 12, 13, 18\}$$

$$E = \text{even numbers of the sample space}$$

$$D = \text{odd numbers of the sample space}$$

(c) List the outcomes of the following events:

$L \cup E$	E^c
$(L \cup E)^c$	$L^c \cap E^c$
L^c	

(d) Construct a Venn diagram which illustrates the fact that, for any two events A and B, it is always true that $(A \cup B)^c = A^c \cap B^c$.

SECTION 6.B
PROBABILITIES OF EVENTS

By the probability of an event which can be repeated over and over again, we mean the proportion of times that the event can be expected to occur relative to the number of times the experiment is repeated. In the case of the probability cubes (dice) discussed in Sec. 6.A, we are talking about a repeatable experiment, rolling the die.

A die is said to be "fair" if each of the six sides has the same probability of being in the upward position after a roll. Because there are six possible outcomes in the sample space $S = \{1, 2, 3, 4, 5, 6\}$, we say that each outcome has probability one-sixth (1/6). In particular, we have recorded in Table 6.1 the facts that $P(\{1\}) = 1/6$, $P(\{2\}) = 1/6$, $P(\{3\}) = 1/6$, $P(\{4\}) = 1/6$, $P(\{5\}) = 1/6$, and $P(\{6\}) = 1/6$, where the symbol $P(\ )$ is pronounced "the probability of." These facts can be interpreted to mean that, in a large number of rolls of the die, we would expect each outcome to occur about one-sixth of the time.

Let's take a look at the following events which we worked with in Sec. 6.A.

$E = \{2, 4, 6\}$ $H = \{5, 6\}$

$L = \{1, 2\}$ $A = \{1, 2, 3, 4\}$

What do we mean by $P(E)$, the probability of E? Well, E is the event that we roll an even number, and since there are three even numbers among the six numbers available, we can expect to roll an even number about three-sixths of the time. For this reason, we can agree that

$$P(E) = \frac{3}{6} = \frac{1}{2}$$

Let's look at $P(E)$ another way. The event E is composed of three distinct outcomes, each of which occurs one-sixth of the time on the average. Therefore the event E will occur that one-sixth of the time that $\{2\}$ occurs, that one-sixth of the

TABLE 6.1
Probabilities for a fair die

Outcome, number of dots facing upward	Proportion of Rolls in Which Outcome Occurs	Probability of Outcome
1	1/6	1/6
2	1/6	1/6
3	1/6	1/6
4	1/6	1/6
5	1/6	1/6
6	1/6	1/6

 ESSENTIALS
OF STATISTICS

time that {4} occurs, and the one-sixth of the time that {6} occurs. From this analysis, we can see that

$$P(E) = P(\{2\}) + P(\{4\}) + P(\{6\}) = \frac{1}{6} + \frac{1}{6} + \frac{1}{6} = \frac{3}{6} = \frac{1}{2}$$

Both the analyses used above are correct methods of calculating the probability of the event E. Similar calculations give us the facts that

$$P(L) = \frac{2}{6} = \frac{1}{3} \qquad\qquad P(H) = \frac{2}{6} = \frac{1}{3} \qquad\qquad P(A) = \frac{4}{6} = \frac{2}{3}$$

Now that we have settled on a method for calculating the probabilities of events such as E, L, H, and A, we turn our attention to finding the probabilities of various combinations of these events.

PROBABILITIES OF INTERSECTIONS

We denote by $E \cap L$ the event that the roll of the die results in a number which is both even and low. Because $E \cap L = \{2\}$, it is reasonable to agree that

$$P(E \cap L) = P(\{2\}) = \frac{1}{6}$$

In a similar manner, we find that

$$P(E \cap A) = P(\{2, 4\}) = \frac{2}{6} = \frac{1}{3}$$

$$P(L \cap A) = P(\{1, 2\}) = P(L) = \frac{1}{3}$$

$$P(L \cap H) = P(\phi) = 0$$

The last assertion that $P(L \cap H) = 0$ can be considered as a statement that there is no chance of having the roll result in a number which is both low and high.

PROBABILITIES OF UNIONS

Now that we know that $P(L \cap H) = 0$, what about $P(L \cup H)$? The event $L \cup H$ is the event that a roll of the die results in either a low or a high (or both, if that were possible) number. As we know, a low number will occur about one-third of the time, and a high number will occur about one-third of the time. It therefore seems reasonable to expect that the goal of obtaining an extreme number (low or high) would be attained about two-thirds of the time.

Translating that statement into mathematical symbols gives the assertion that

$$P(L \cup H) = P(L) + P(H) = \frac{1}{3} + \frac{1}{3} = \frac{2}{3}$$

This estimate is corroborated by the following calculations, using the direct method of counting outcomes:

$$P(L \cup H) = P(\{1, 2, 5, 6\}) = \frac{4}{6} = \frac{2}{3}$$

Suppose we apply the same method of analysis to the problem of calculating $P(E \cup A)$. We know that an even number occurs about one-half of the time, while a not-too-high number occurs about two-thirds of the time. It definitely cannot be true, however, that

$$P(E \cup A) = P(E) + P(A) = \frac{1}{2} + \frac{2}{3} = \frac{3}{6} + \frac{4}{6} = \frac{7}{6}$$

because we would be saying that $E \cup A$ could be expected to occur about seven out of every six times on the average, which is more than 100% of the time. (Probabilities can never exceed one, for, if they did, the events involved would have to be occurring more than 100% of the time.) What, then, went wrong with the statement that $P(E \cup A) = P(E) + P(A)$? If we look at the Venn diagram in Fig. 6.7, we observe that the union $E \cup A$ consists of five outcomes. It therefore seems reasonable to believe that $P(E \cup A) = 5/6$. This belief is substantiated by the direct method of counting outcomes, namely,

$$P(E \cup A) = P(\{1, 2, 3, 4, 6\}) = \frac{5}{6}$$

FIGURE 6.7
Probability of the union of events.

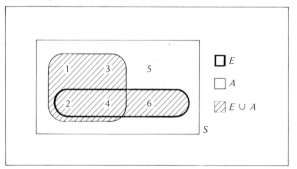

How, then, did we manage to come up with the allegation that $P(E \cup A) = 7/6$? If we look closely at the statement

$$P(E \cup A) = P(E) + P(A) = \frac{3}{6} + \frac{4}{6} = \frac{7}{6}$$

we will see, hidden inside it, the following statement:

$$P(E \cup A) = P(E) + P(A)$$
$$= P(\{2, 4, 6\}) + P(\{1, 2, 3, 4\})$$
$$= \frac{3}{6} + \frac{4}{6} = \frac{7}{6}$$

Take a look at what we have done. We have blatantly counted the outcomes 2 and 4 twice: once as part of E and once as part of A. In short, while computing the probability of $E \cup A = \{1, 2, 3, 4, 6\}$, we have counted the outcomes 1, 3, and 6 once and the outcomes 2 and 4 twice. Therefore, we have implicitly been assuming that $E \cup A$ contains *seven* outcomes, namely, $\{1, 2, 2, 3, 4, 4, 6\}$, where in fact it contains only *five*.

Now that we have discovered the mistake, how do we rectify it? How do we correct the erroneous statement that $P(E \cup A) = P(E) + P(A)$? Well, when we were using $P(E) + P(A)$ to calculate $P(E \cup A)$, we were counting twice all those outcomes which are in both E and A, namely, all those outcomes in $E \cap A = \{2, 4\}$. But we want to count these outcomes only once. We have therefore counted the outcomes in $E \cap A$ once too often. We can correct the mistake by subtracting the overcount of 2 and 4, an amount equal to $P(E \cap A) = P(\{2, 4\}) = 2/6 = 1/3$. We can therefore write that

$$P(E \cup A) = P(E) + P(A) - P(A \cap E)$$
$$= \frac{3}{6} + \frac{4}{6} - \frac{2}{6} = \frac{5}{6}$$

which is the correct probability. We can summarize this development in the following general rule: If C and D are any two events, then

★ $\quad P(C \cup D) = P(C) + P(D) - P(C \cap D)$ $\qquad\qquad$ (6.2)

Now that we have established the general rule, let's go back a page or two and try to figure out how we managed to get away with the statement that

$$P(L \cup H) = P(L) + P(H) = \frac{1}{3} + \frac{1}{3} = \frac{2}{3}$$

If we apply the general rule to $L \cup H$, we should really be writing that

$$P(L \cup H) = P(L) + P(H) - P(L \cap H)$$

What about $P(L \cap H)$? Well, as it turns out, $L \cap H = \phi$ because none of the possible outcomes can be found in both L and H. Therefore $P(L \cap H) = P(\phi) = 0$. It follows that

$$P(L \cup H) = P(L) + P(H) - P(L \cap H)$$
$$= \frac{1}{3} + \frac{1}{3} - 0 = \frac{2}{3}$$

and so the missing term $P(L \cap H)$ had no effect on the calculation of $P(L \cup H)$. We did not even realize that we were leaving it out.

We now return to a further discussion of the important relationship (6.1): If E and F are any two events, we discovered that

$$E = (E \cap F) \cup (E \cap F^c)$$

What can we say about $P(E) = P([(E \cap F) \cup (E \cap F^c)])$? Well, notice from Fig. 6.6 that $E \cap F$ is contained entirely inside of F, while $E \cap F^c$ is contained entirely inside of F^c. If there were some outcomes in the intersection $(E \cap F) \cap (E \cap F^c)$, they would have to be in $E \cap F$ (therefore in F) and also in $E \cap F^c$ (therefore in F^c). Those outcomes would then have to be in both F and F^c. But there cannot be any outcomes which are simultaneously in both F and F^c, for F^c consists only of those outcomes which are not in F. Therefore, there cannot be any outcomes which are in $(E \cap F) \cap (E \cap F^c)$. It follows that $(E \cap F) \cap (E \cap F^c) = \phi$, and so by analogy with $P(L \cup H)$,

$$P([(E \cap F) \cup (E \cap F^c)]) = P(E \cap F) + P(E \cap F^c)$$

We can therefore write that

★ $$P(E) = P(E \cap F) + P(E \cap F^c) \tag{6.3}$$

EXERCISES 6.B

1   Consider the experiment of tossing a pair of fair dice, one of them red and the other green. Because there are six possible ways for each of the dice to turn up (namely, 1, 2, 3, 4, 5, or 6), there are 36 possible ways for the pair to turn up. For example, we can get 1 on the red and 3 on the green, or 6 on the red and 4 on the green, or 3 on the red and 1 on the green, or 5 on the red and

5 on the green, etc. Since the dice are fair, each of these 36 possible ways is equally likely and so has probability 1/36 of turning up. To calculate the probability of rolling a total of 4 dots on the two dice together, we notice that 3 of the 36 possibilities have a total of 4 dots, namely, (1) 1 on the red and 3 on the green, (2) 2 on the red and 2 on the green, and (3) 3 on the red and 1 on the green. Therefore the probability of rolling a 4 with a pair of fair dice is 3/36.

(a) Find the probability of rolling a 7.
(b) Find the probability of rolling an 11.
(c) Find the probability of the event $W = \{7, 11\}$.
(d) Find the probability of the event $L = \{2, 3, 12\}$.
(e) Find the probability of the event $E = \{2, 4, 6, 8, 10, 12\}$.
(f) Find the probability of the event $D = \{3, 5, 7, 9, 11\}$.
(g) Find the probability $P(L \cap D)$.
(h) Find $P(L \cup D)$.
(i) Find $P(W \cap L)$.
(j) Find $P(W \cup L)$.
(k) Find $P(E \cap D)$.
(l) Find $P(E \cup D)$.

2 If A and B are two events such that $P(A) = 1/2$, $P(B) = 2/3$, and $P(A \cap B) = 1/3$, then determine the following probabilities:

(a) $P(A^c)$ (c) $P(A^c \cap B)$
(b) $P(B^c)$ (d) $P(A^c \cap B^c)$

3 If Q and R are two events such that $P(Q) = 1/2$, $P(R) = 3/8$, and $P(Q \cap R) = 1/4$, then determine the following probabilities:

(a) $P(Q^c)$ (c) $P(Q^c \cap R)$
(b) $P(R^c)$ (d) $P(Q^c \cap R^c)$

4 Explain why it is impossible to have two events U and V such that $P(U) = 2/3$, $P(V) = 4/5$, and $P(U \cap V) = 1/4$.

5 Explain why it is impossible to have two events T and W such that $P(T) = 1/8$, $P(W) = 1/5$, and $P(T \cup W) = 1/2$.

SECTION 6.C
UPDATING PROBABILITIES OF EVENTS

In most questions in applied fields which involve uncertainty, the researcher or manager is called upon to provide an estimate of the probability of some uncertain event. Often the event of interest is not the sort that can be repeated over and over again. Therefore, it is not possible to estimate its probability by finding the relative frequency of its occurrence. The best procedure for finding the probability of the event would then be to gather as many of the relevant facts about the situation as possible, to try to determine what effect each of the facts has upon the probability, and then to come up with an estimate of the desired probability.

As an example, suppose we want to estimate the probability of the event

I = event that a particular individual will sustain a very serious injury within the week

If actuarial statistics indicate that about one out of every million individuals sustains a very serious injury every week, in probabilistic terminology this can be stated as

$P(I)$ = .000001

After making this estimate of $P(I)$, suppose that new information comes in about the particular individual under discussion. Say, for example, that we have the information

P = the individual is a licensed pilot of small airplanes

Since the injury rates for pilots of small planes are somewhat higher than those of the general public, perhaps

$P(I|P)$ = .0001

namely, one out of ten thousand. The symbol $P(I|P)$ is pronounced "the probability of I given P," and it means the probability of I updated so as to reflect the new information contained in statement P. In particular $P(I|P)$ is the probability that an individual will sustain a very serious injury within the week if that individual is a licensed pilot of small airplanes. Updated probabilities such as $P(I|P)$ are technically referred to as "conditional probabilities," for they measure the probabilities of events updated to take changing conditions into account.

In many cases of applied interest, new information about a changing situation continues to flow in, and the probabilities must be continually updated. Suppose the following fact about our current problem becomes known:

C = the individual regularly pilots a crop-dusting plane

Because piloting a crop-dusting plane is somewhat more hazardous than piloting other small planes (because crop-dusting planes commonly fly about 10 to 20 feet above the ground, with trees, wires, etc., in their paths), it would be reasonable to update the probability of injury to

$P(I|C)$ = .001

or one chance out of a thousand. More information flows in, with each piece of information having an effect on the probability of injury. The bits of information and their effects on the probability of injury are listed in Table 6.2. Some pieces

TABLE 6.2
Effect of new information on the probability of injury

I = event that an individual will sustain a very serious injury within the week
$P(I)$ = .000001 (at 12:00 noon)

Time of Receipt of Information, P.M.	Gist of New Information	Symbol	Updated Probability
1:00	He is licensed to pilot small planes.	P	$P(I \mid P)$ = .0001
2:00	He regularly pilots crop dusters.	C	$P(I \mid C)$ = .001
3:00	He fell out of his plane while crop-dusting.	F	$P(I \mid F)$ = .99
3:30	There was an extremely large haystack below the plane.	H	$P(I \mid H)$ = .05
3:45	There was a bull sleeping in the haystack.	B	$P(I \mid B)$ = .75
3:47	He missed the bull.	MB	$P(I \mid MB)$ = .05
3:48	He missed the haystack.	MH	$P(I \mid MH)$ = .99

of information tend to increase the probability of I, while other pieces tend to decrease it. Table 6.2 shows that the conditional probability fluctuates up and down as conditions change. The events $P, C, F, B,$ and MH tend to exert upward pressure on the probability of I, while the events H and MB tend to exert downward pressure.

To illustrate the calculation of more concrete conditional probabilities, let's return to the probability cube example of the last two sections and the following four events:

$$E = \{2, 4, 6\} \quad P(E) = 1/2$$

$$L = \{1, 2\} \quad P(L) = 1/3$$

$$H = \{5, 6\} \quad P(H) = 1/3$$

$$A = \{1, 2, 3, 4\} \quad P(A) = 2/3$$

We know that the probability of L is 1/3. Suppose that we have received new information to the effect that the roll of the die resulted in the occurrence of the event A. What is the updated probability of L, in view of the new information that A has occurred? We come up with the numerical value of $P(L|A)$ by the following reasoning: The fact that A has occurred means that the number of dots on the upward side of the die was either 1, 2, 3, or 4. The possibility that 5 or 6 might have been rolled was excluded by the occurrence of event A. We have no information asserting that any of the numbers 1, 2, 3, or 4 was more likely to occur than any of the others; so the nature of the probability cube requires that the chances of each of these numbers be updated equally to 1/4. This is because there are now only four possible outcomes, taking the new information into account that the outcomes 5 and 6 definitely did not occur. Since the only possible outcomes were 1, 2, 3, and 4, the updated probability of $L = \{1, 2\}$ is then

$$P(L|A) = \frac{2}{4} = \frac{1}{2}$$

because L includes two of the four possible outcomes. While the original probability of L was 1/3, we see that the conditional probability of L, given A, is 1/2. Therefore the occurrence of A has *increased* the probability of L because $P(L|A) > P(L)$. We say that the event A is *favorable* to the event L.

If, instead of the occurrence of A, we were given the information that the event H had occurred, we would be interested in calculating $P(L|H)$, the updated probability of L. The information that H has occurred automatically means that the number rolled on the die was definitely either a 5 or a 6. There is no longer any possibility that a 1, a 2, a 3, or a 4 was rolled. But $L = \{1, 2\}$, so that there is no chance that event L could've occurred. It follows that

$$P(L|H) = 0$$

Because the original probability of L was 1/3 and the conditional probability of L given H is 0, we can see that $P(L|H) < P(L)$. We say that the event H is *unfavorable* to the event L, for the occurrence of H has *decreased* (quite substantially in this case) the probability of L.

Suppose, finally, that the information we were given was that the event E had occurred. The occurrence of E would mean that the actual number rolled was either 2, 4, or 6. The only possible outcome which would therefore result in the occurrence of L would be the 2. There is no chance that L's outcome 1 would occur, for the occurrence of E specifically excludes the occurrence of 1, 3, and 5. Therefore the numbers 2, 4, and 6 would each be assigned an updated probability of 1/3, since those are now the only three possible outcomes. The occurrence of 2 would imply the occurrence of L, while the occurrence of 4 or 6 would not. Therefore the updated probability of L, in view of the information that E has occurred, is

$$P(L|E) = \frac{1}{3}$$

because L includes one of the three possible outcomes 2, 4, and 6. We note something unusual here: The original probability of L was 1/3, while the conditional probability of L given E also turned out to be 1/3. Therefore, it seems that the occurrence of E did *not affect* in any way the chances of L, because $P(L|E) = P(L)$. We describe this situation by saying that the event L is *independent* of the event E.

We summarize the types of relationships between the original and the updated probabilities in Table 6.3.

TABLE 6.3
Conditional probability relationships

Verbal Expression	Mathematical Expression
C is favorable to D.	$P(D\mid C) > P(D)$
C is unfavorable to D.	$P(D\mid C) < P(D)$
D is independent of C.	$P(D\mid C) = P(D)$

In many cases of applied interest, it will be useful to have a formula for computing conditional probabilities in terms of original probabilities. In order to develop such a formula, let's sit down and try to figure out what the conditional probability of C given D really is. Consider the Venn diagram on the left side of Fig. 6.8. The original probability of C can be viewed as the proportion of the sample space S that is occupied by C. If C is a large part of S, the probability of C will be high, while if C is a small part of S, the probability of C will be small. What happens to C after D occurs? This situation is illustrated on the right side of Fig. 6.8. The information that D has occurred has the effect of reducing the sample space to D because the outcomes not in D can no longer be considered as possible outcomes. The probability of C in this situation is then the proportion of the new sample space D that is occupied by C. The part of D that is occupied by C is the set $C \cap D$. Therefore, if $C \cap D$ is a large part of D, the conditional probability of C given D will be large. If $C \cap D$ is a small part of D, the conditional probability of C given D will be small. To express the conditional probability as the proportion of D occupied by $C \cap D$, we have the following formula for conditional probability

$$\bigstar \quad P(C\mid D) = \frac{P(C \cap D)}{P(D)} \tag{6.4}$$

Before illustrating the use of this formula in applied work, let's use it to recalculate the conditional probabilities that we found directly for the roll of a die. We

FIGURE 6.8
The condition probability of C, given D.

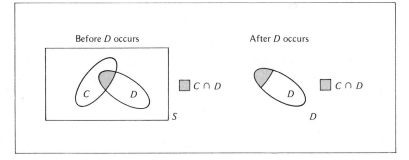

have

$$P(L|A) = \frac{P(L \cap A)}{P(A)} = \frac{P(\{1, 2\})}{P(\{1, 2, 3, 4\})} = \frac{1/3}{2/3} = \frac{1}{2}$$

$$P(L|H) = \frac{P(L \cap H)}{P(H)} = \frac{P(\phi)}{P(\{5, 6\})} = \frac{0}{1/3} = 0$$

$$P(L|E) = \frac{P(L \cap E)}{P(E)} = \frac{P(\{2\})}{P(\{2, 4, 6\})} = \frac{1/6}{1/2} = \frac{2}{6} = \frac{1}{3}$$

Naturally, these results are exactly the same as we had obtained earlier.

If in formula (6.4) for conditional probability we multiply both sides of the equation through by $P(D)$, we see that

$$P(C|D)\, P(D) = \frac{P(C \cap D)}{P(D)}\ P(D) = P(C \cap D)$$

as the two $P(D)$'s cancel out. The resulting expression

★ $P(C \cap D) = P(C|D)\, P(D)$ (6.5)

is very useful in applications, perhaps just as useful as the formula for conditional probability itself.

Example 6.1 A Bond Referendum An elementary school bond referendum which requires a majority vote to pass is put before the voters in a school district where 30% of the voters have children in elementary school and 70% do not. Polls indicate that 90% of those with children in elementary school will vote for the bond issue, while only 20% of those without children in elementary school will do so. If the polls are right, will the bond issue pass?

SOLUTION   We need to compactly label some events. The important ones are

$F =$ event that voter voted for bond issue

$E =$ event that voter had children in elementary school

In the language of these events we know that $P(E) = .30$ and $P(E^c) = .70$ for the school district under study. The polls indicate that $P(F|E) = .90$ and $P(F|E^c) = .20$. What we want to know is $P(F)$, the proportion of voters who voted for the bond issue. By formula (6.3), we know that

$$P(F) = P(F \cap E) + P(F \cap E^c)$$

Now using formula (6.5) we can calculate each of the terms $P(F \cap E)$ and $P(F \cap E^c)$. We have that

$$P(F \cap E) = P(F|E)\ P(E) = (.90)(.30) = .27$$

and

$$P(F \cap E^c) = P(F|E^c)\ P(E^c) = (.20)(.70) = .14$$

It follows that

$$P(F) = P(F \cap E) + P(F \cap E^c) = .27 + .14 = .41$$

and so the indications are that the bond issue will get 41% of the votes, which is short of the majority required for passage.

We can use a picture called a "tree diagram" to illustrate the necessary calculations. The diagram appears in Fig. 6.9.

Example 6.2 Automobile Insurance An insurance company issues three types of automobile policies: Type G for good risks, Type M for moderate risks, and Type B for bad risks. The company's clients are 20% Type G, 40% Type M, and

FIGURE 6.9
Tree diagram of the bond referendum example.

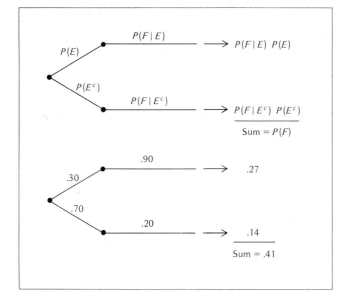

40% Type *B*. Accident statistics reveal that a Type *G* driver has probability .01 of causing an accident in a 12-month period, a Type *M* driver has probability .02, and a Type *B* driver has probability .08. What proportion of the company's clients will cause an accident in the next 12 months?

SOLUTION   The pertinent events are as follows:

A = event that client will cause accident in next 12 months

G = event that client is Type *G* driver

M = event that client is Type *M* driver

B = event that client is Type *B* driver

The answer to the question is $P(A)$, the proportion of clients who will cause an accident in the next 12 months. From an extension of formula (6.3), we see that

$$P(A) = P(A \cap G) + P(A \cap M) + P(A \cap B)$$

Applying the conditional probability expression (6.5), we get that

$$P(A) = P(A|G)\,P(G) + P(A|M)\,P(M) + P(A|B)\,P(B)$$

FIGURE 6.10
Tree diagram of the accident insurance example.

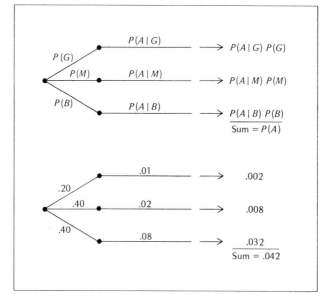

Because the information given in the example translates into the following facts:

$P(G) = .20$	$P(A\|G) = .01$
$P(M) = .40$	$P(A\|M) = .02$
$P(B) = .40$	$P(A\|B) = .08$

we can calculate

$$P(A) = (.01)(.20) + (.02)(.40) + (.08)(.40)$$
$$= .002 + .008 + .032 = .042$$

so that 4.2% of the company's clients can be expected to cause an accident in the next 12 months. The calculation is illustrated by a tree diagram in Fig. 6.10.

EXERCISES 6.C

1 Consider again the experiment of tossing a pair of fair dice, one of them red and the other green. (This situation was previously discussed in Exercises 6.A.1 and 6.B.1.)
   (a) Defining the events $W = \{7, 11\}$ and $D = \{3, 5, 7, 9, 11\}$, calculate the probability $P(W|D)$.
   (b) Defining the events $L = \{2, 3, 12\}$ and $E = \{2, 4, 6, 8, 10, 12\}$, calculate the probability $P(L|E)$.
   (c) Find $P(D|W)$.
   (d) Find $P(E|L)$.
   (e) Find $P(W|E)$.
   (f) Find $P(L|D)$.
2 A jar contains eight green balls and two purple balls. Balls are drawn from the jar according to the following scheme: Whenever a green ball is drawn, it is replaced by a new green ball before the next draw, but whenever a purple ball is drawn, it is replaced by three new purple balls before the next draw. For example, if a purple ball is drawn on the first draw, then there will be eight green balls and four purple balls in the jar awaiting the second draw.
   (a) Find the probability that the first ball drawn is green.
   (b) If the first ball drawn is green, find the probability that the second ball drawn will also be green.
   (c) Find the probability that the first ball drawn is purple.
   (d) If the first ball drawn is purple, find the probability that the second ball drawn will be green.
   (e) Find the probability that the second ball drawn will be green.
3 In a learning experiment on a T maze, a rat is equally likely to turn left (receiving an electric shock) or right (receiving food) on his first time

through. If he received food the first time, he has probability 3/5 of turning right again the second time. If, on the other hand, he received a shock the first time, he has probability 4/5 of turning right the second time.

(a) What is the probability that the rat will turn right both times?

(b) What is the probability that the rat will turn left on the first try but right on the second?

(c) What is the probability that the rat will turn left both times?

(d) What is the probability that the rat will receive food on the second try?

4 One psychologist has studied the incidence of grimacing and rheumatic disease in schizophrenic patients. In a study of 2000 such patients, he found that 100 had rheumatic disease, and of those 100, 40 also had grimacing. Another 360 had grimacing but no rheumatic disease.

(a) What was the probability that a patient had grimacing?

(b) What was the conditional probability that a patient had grimacing given that the patient had rheumatic disease?

(c) Is grimacing independent of rheumatic disease in schizophrenic patients?

5 On the average, about 5% of the patients assigned to a state hospital which specializes in the treatment of respiratory diseases have tuberculosis. All persons entering this hospital are given chest x-rays, and past data indicate that 90% of those with tuberculosis have a positive x-ray (substantiating presence of the disease), while only 2% of those without tuberculosis have a positive x-ray. How many patients have positive x-rays?

6 A national sociological survey indicates that 15% of all adults are college graduates, while 85% are not. Furthermore, 80% of all college graduates own their own homes, while only 30% of nongraduates are homeowners. What proportion of adults are homeowners?

7 Patients manifesting symptoms of certain psychological disorders are often tested for the Archimedes spiral aftereffect. Of the patients tested, accumulated data indicate that 40% have a functional illness, 30% are brain-damaged, and the other 30% have neither of these disorders. Of known functionally ill patients, 80% have the aftereffect when tested, while only 30% of brain-damaged patients and 10% of the others have the aftereffect. What percentage of patients have the aftereffect?

8 If interest rates are headed downward, the probability is 70% that the stock market will go up. If interest rates are likely to go up, the chances are 80% that the market will go down. If interest rates are remaining stable, the chances are 60% that the market will go up. Past history indicates that interest rates go down 30% of the time, go up 20% of the time, and remain stable 50% of the time. What percentage of the time does the stock market rise?

9 Of those who voted in a recent election, 10% were college graduates, 50% were high-school graduates, the remaining 40% were grade-school graduates. Alphonse, the victorious candidate, received 80% of the college-graduate vote, 70% of the high-school-graduate vote, and 40% of the grade-school-graduate vote. What percentage of voters voted for Alphonse?

10 Eighty percent of used car and major appliance buyers are good credit risks. Seventy percent of good credit risks have a charge account in at least one department store, while 40% of bad credit risks have a charge account. What percentage of prospective buyers have charge accounts?

SECTION 6.D
INDEPENDENT EVENTS

When $P(C|D)$ is different from $P(C)$, the knowledge of the occurrence of D has had an influence on our view of the likelihood of the occurrence of C. We can describe this situation by saying that the event C "depends on" the event D, in the sense that the occurrence of D affects the probability of C. If $P(C|D)$ is exactly the same, numerically speaking, as $P(C)$, then the occurrence of D has had no influence on the probability of C. We therefore say that the event C is "independent" of the event D.

The concept of independent events is often mistaken for the concept of disjoint events. In fact, however, independence and disjointness are extreme opposites of each other. If the events C and D are disjoint, then $C \cap D = \phi$, so that by the formula for conditional probability,

$$P(C|D) = \frac{P(C \cap D)}{P(D)} = \frac{P(\phi)}{P(D)} = \frac{0}{P(D)} = 0$$

no matter what the original probability of C. What this means is that, even if $P(C)$ is very high, the occurrence of D eliminates all chance of C's occurrence because C and D are mutually exclusive. Therefore, C depends very heavily on D, since the conditional probability of C given D is very small compared with the original probability of C.

A very useful fact about independent events is that, if C is independent of D, then D is automatically independent of C. Specifically, if the occurrence of C does not affect the probability of D, then the occurrence of D would not affect the probability of C. Mathematically, this fact can be expressed as follows: If $P(C|D) = P(C)$, then $P(D|C) = P(D)$. The justification of this fact involves the use of the formulas for conditional probability (6.4) and (6.5). If $P(C|D) = P(C)$, then

$$P(D|C) = \frac{P(D \cap C)}{P(C)} = \frac{P(C \cap D)}{P(C)} = \frac{P(C|D) \, P(D)}{P(C)}$$

$$= \frac{P(C) \, P(D)}{P(C)} = P(D)$$

where we have replaced $P(C|D)$ by its equal $P(C)$. Because of this symmetry of meaning of the word "independence," we can refer to C and D as a pair of independent events.

There is an algebraic expression of the meaning of independence which reflects this symmetry. The independence of C and D means that $P(C) = P(C|D)$, so that

$$P(C) = \frac{P(C \cap D)}{P(D)}$$

which, by cross-multiplication, becomes

$$P(C) \, P(D) = P(C \cap D)$$

It then follows that the events C and D are independent if and only if

★ $P(C \cap D) = P(C) \, P(D)$ (6.6)

The above expression of independence is the one most useful in the solution of applied problems. It forms the basic theoretical foundation for the chi-square test of independence to be discussed in the next section.

In dealing with more than two independent events, we need an expanded version of the above equation. We will say that n events $C_1, C_2, \ldots, C_n$ are independent if every event is independent not only of every other event but also of every possible combination of all other events. It turns out that

★ $P(C_1 \cap C_2 \cap \cdots \cap C_n) = P(C_1) \, P(C_2) \cdots P(C_n)$ (6.7)

if the events $C_1, C_2, \ldots, C_n$ are independent. This means that the probability that all the events will occur is obtained by multiplying together all their original probabilities.

Example 6.3 Coin Tossing   What is the probability that a fair coin will fall heads on ten consecutive tosses?

SOLUTION   Denote by H_k the event that the coin falls heads on the kth toss, i.e.,

H_1 = event that coin falls heads on first toss

H_2 = event that coin falls heads on second toss

$\vdots$

H_{10} = event that coin falls heads on tenth toss

Then, because the coin does not remember which side it landed on each time, the events $H_1, H_2, H_3, H_4, H_5, H_6, H_7, H_8, H_9$, and H_{10} are independent. Each of these events has probability 1/2 because the coin is fair, so that heads and tails are equally likely. Therefore the probability of heads on all ten tosses is

$$P(H_1 \cap H_2 \cap H_3 \cap H_4 \cap H_5 \cap H_6 \cap H_7 \cap H_8 \cap H_9 \cap H_{10})$$
$$= P(H_1)P(H_2)P(H_3)P(H_4)P(H_5)P(H_6)P(H_7)P(H_8)P(H_9)P(H_{10})$$
$$= \left(\frac{1}{2}\right)\left(\frac{1}{2}\right)\left(\frac{1}{2}\right)\left(\frac{1}{2}\right)\left(\frac{1}{2}\right)\left(\frac{1}{2}\right)\left(\frac{1}{2}\right)\left(\frac{1}{2}\right)\left(\frac{1}{2}\right)\left(\frac{1}{2}\right)$$
$$= \frac{1}{1024} = .000977$$

using the expanded version (6.7) of the independence formula. Therefore the chances of tossing ten heads in a row with a fair coin are slightly less than one in a thousand. (Because coin tossing is one of the prime examples of a process which generates binomially distributed data, we could have read the desired probability directly from Table A.2 which has itself been constructed using the procedure we have just carried out.)

Example 6.4 A Political Poll Candidates Able and Baker, running for the same office, are facing an electorate of several million persons. A national poll claims that two out of every three voters favor Able over Baker. Suppose we want to test the poll's validity in our district of several thousand persons. We choose a random sample of five voters and we ask ourselves the following question: "If it were really true that two out of three favored Able, what is the probability that in our sample of five voters, only one of the five would favor Able?"

SOLUTION   The relevant events are the following:

A_1 = event that 1st voter in sample favors Able

A_2 = event that 2d voter in sample favors Able

$\vdots$

B_1 = event that 1st voter in sample favors Baker

$\vdots$

B_5 = event that 5th voter in sample favors Baker

If it were really true that two out of three voters favor Able, the probabilities would be as follows:

$$P(A_1) = P(A_2) = P(A_3) = P(A_4) = P(A_5) = \frac{2}{3}$$

and

$$P(B_1) = P(B_2) = P(B_3) = P(B_4) = P(B_5) = \frac{1}{3}$$

We are interested in $P(E)$, the probability that exactly one of the five voters selected favors Able. Because of the large number of voters from which the five are to be selected, it is reasonable to consider the five voters selected as independent of each other. The following events together constitute E, the event that exactly one of the five voters favors Able:

$A_1^* = A_1 \cap B_2 \cap B_3 \cap B_4 \cap B_5$

   = event that 1st voter favors Able while other four favor Baker

$A_2^* = B_1 \cap A_2 \cap B_3 \cap B_4 \cap B_5$

   = event that 2d voter favors Able while other four favor Baker

$\vdots$

$A_5^* = B_1 \cap B_2 \cap B_3 \cap B_4 \cap A_5$

   = event that 5th voter favors Able while other four favor Baker

Because $E = A_1^* \cup A_2^* \cup A_3^* \cup A_4^* \cup A_5^*$, where the latter five events are disjoint, an expanded version of formula (6.2) implies that $P(E) = P(A_1^*) + P(A_2^*) + P(A_3^*) + P(A_4^*) + P(A_5^*)$. It remains now only to compute the probabilities of the events A_k^*. Because the voters are assumed to be independent of each other, we have, by the expanded formula (6.7) for independent events, that

$P(A_1^*) = P(A_1)\, P(B_2)\, P(B_3)\, P(B_4)\, P(B_5)$

$$= \left(\frac{2}{3}\right)\left(\frac{1}{3}\right)\left(\frac{1}{3}\right)\left(\frac{1}{3}\right)\left(\frac{1}{3}\right) = \frac{2}{243} = .00823$$

$$P(A_2^*) = \left(\frac{1}{3}\right)\left(\frac{2}{3}\right)\left(\frac{1}{3}\right)\left(\frac{1}{3}\right)\left(\frac{1}{3}\right) = \frac{2}{243} = .00823$$

$\vdots$

$$P(A_5^*) = \left(\frac{1}{3}\right)\left(\frac{1}{3}\right)\left(\frac{1}{3}\right)\left(\frac{1}{3}\right)\left(\frac{2}{3}\right) = \frac{2}{243} = .00823$$

The five events A_1^*, A_2^*, A_3^*, A_4^*, and A_5^* each have probability .00823. It follows that the probability that exactly one of the five voters favors Able is

$P(E) = .00823 + .00823 + .00823 + .00823 + .00823$

   $= .04115 \approx 4\%$

Therefore, if two out of every three voters really favored Able, the chances are only about 4% that exactly one of five randomly selected voters would favor Able. If we were to choose a sample resulting in only one of the five favoring Able, it would then be reasonable to assume that in our district it is probably not true that two out of three favor Able.

EXERCISES 6.D

1 Suppose we toss a fair coin six times. Determine the following probabilities and compare your answers with the appropriate numbers listed in Table A.2 for the binomial distribution with parameters $n = 6$ and $p = 1/2$:
   (a) The probability that none of the six tosses results in a head.
   (b) The probability that exactly one of the six tosses results in a head.
   (c) The probability that exactly two of the six tosses result in heads.
   (d) The probability that exactly three of the six tosses result in heads.
   (e) The probability that exactly four of the six tosses result in heads.
   (f) The probability that exactly five of the six tosses result in heads.
   (g) The probability that all six tosses result in heads.
   (h) Show that your answers a through g add up to one.

2 Suppose we roll a fair die six times. Determine the following probabilities and compare your answers with the appropriate formula for the binomial distribution with parameters $n = 6$ and $p = 1/6$:
   (a) The probability that none of the six rolls results in a five.
   (b) The probability that exactly one of the six rolls results in a five.
   (c) The probability that exactly two of the six rolls result in fives.
   (d) The probability that exactly three of the six rolls result in fives.
   (e) The probability that exactly four of the six rolls result in fives.
   (f) The probability that exactly five of the six rolls result in fives.
   (g) The probability that all six rolls result in fives.
   (h) Show that your answers a through g add up to one.

3 A community within a small urban area has two ambulances available, one at the north end of town and the other at the south end. The two ambulances operate independently of each other, but due to differences in mechanical condition and traffic patterns, they have probabilities .9 and .5, respectively, of arriving at an accident scene within 10 minutes.
   (a) If an accident requires both ambulances, what is the probability that they both arrive within 10 minutes?
   (b) What is the probability that the one from the north end arrives within 10 minutes, but the one from the south end does not?
   (c) What is the probability that both take longer than 10 minutes to arrive on the scene?

4 A small machine shop is considering purchase of an oscilloscope with four vital parts, all of which work properly or fail independently of the others. The machine fails to work properly if two or more of these four parts fail. The shop's tests have yielded the following data on operating probabilities:

Part	Probability of Working Properly
A	.9
B	.9
C	.8
D	.5

(a) What is the probability that all four parts work properly simultaneously?
(b) What is the probability that A fails, while B, C, and D all work properly?
(c) What is the probability that D fails, while A, B, and C all work properly?
(d) What is the probability that the oscilloscope works properly?

5 Suppose it is really true that 60% of the voters are for repeal of a certain local tax law. If we were to select a random sample of seven voters, what would be the probability that a majority of the sample would be against repeal, thus making it appear that the antirepeal forces are in the lead? Compare your solution using probability theory with that obtained in Example 2.2 of Chap. 2.

6 The probability that a new type of communications satellite will be successfully launched into orbit is .8.
(a) If launchings are continually attempted until one is successful, what is the probability that no more than two attempts will be necessary?
(b) If launchings are attempted until two are successful, what is the probability that no more than four attempts will be necessary?
(c) Which probability is larger? Can you give a verbal explanation of the reasons why this should be so?

SECTION 6.E
THE CHI-SQUARE TEST OF INDEPENDENCE

A survey was conducted to evaluate the relative effectiveness of two allergy remedies which had been administered in a small community. The treatments were provided in the spring, free of charge to those wishing to avail themselves of them. Some people received the first treatment, an allergy shot, while others preferred the second treatment, an allergy capsule, and the remaining persons received no treatment. A random sample of 1000 local inhabitants the following fall yielded the results presented in Table 6.4. The figures in Table 6.4 are interpreted as follows: Of the 1000 people surveyed, 350 had received no treatment. Of those 350, 44 had manifested severe allergy symptoms while the remaining 306 had not. Those in the sample who had taken the allergy capsule numbered 150, and of these, 19 had severe symptoms while 131 escaped them. Finally, 500 persons out of the 1000 had been treated with an allergy shot, with 37 reporting severe symptoms and 463 managing to avoid them.

The question we will try to answer in this section is the following: "Is the successful avoidance of severe allergy symptoms independent of the receipt of treatment?"

TABLE 6.4
Contingency table of allergy treatment data

Treatment Record \ Allergy Record	Severe Symptoms	Mild or No Symptoms	Row Totals
No treatment	44	306	350
Allergy capsule	19	131	150
Allergy shot	37	463	500
Column totals	100	900	1000

For the theoretical structure upon which the practical techniques of this section are built, we turn back to Sec. 6.D. According to Eq. (6.6), the events A and B are independent if and only if $P(A \cap B) = P(A)P(B)$. Otherwise the events are said to be dependent. Our goal will be to develop a statistical test, called the "chi-square test of independence," which will tell us (on the basis of data displayed in a "contingency table" of the sort illustrated in Table 6.4) when two characteristics are independent and when they are dependent. (The Greek letter *chi* is pronounced "kye.")

The basic principle underlying the chi-square test is as follows: We calculate a hypothetical contingency table, usually called the "expected contingency table," showing how the data would have turned out if avoidance of severe symptoms were independent of receipt of treatment. Then we compare this expected contingency table with the actual contingency table showing how the data really turned out. If the actual table looks very similar to the expected table, we would conclude that the two characteristics (namely, avoidance of severe symptoms and receipt of treatment) are probably independent. If, on the other hand, the actual table looks quite different from the expected table, we would have to say that the two characteristics are probably not independent.

We will introduce the chi-square statistic as a measure of the difference between the two contingency tables. We will reject the hypothesis of independence if the chi-square value exceeds a corresponding number in the chi-square table, Table A.5 of the Appendix. The chi-square statistic measures the difference between two contingency tables in a manner analogous to the way the two-sample t statistic measures the difference between two sample means.

We now proceed to the calculation of the expected contingency table of the allergy treatment data of Table 6.4. We are dealing with the following five events:

Treatment Events

NT = event that person received no treatment

AC = event that person received allergy capsule

AS = event that person received allergy shot

Symptom Events

SS = event that person reported severe symptoms

MS = event that person reported mild or no symptoms

From Table 6.4 we can find the probabilities of these events:

Treatment Events

$P(NT) = 350/1000$ because 350 out of 1000 received no treatment

$P(AC) = 150/1000$ because 150 out of 1000 received capsule

$P(AS) = 500/1000$ because 500 out of 1000 received shot

Symptom Events

$P(SS) = 100/1000$ because 100 out of 1000 reported severe symptoms

$P(MS) = 900/1000$ because 900 out of 1000 reported mild or no symptoms

From the actual contingency table on the left side of Table 6.5, we see that the event $NT \cap SS$ consists of 44 persons. This means that there were exactly 44 persons who both received no treatment and also exhibited severe symptoms. Let's now compute the expected number of persons in the event $NT \cap SS$ in the case of independence. If NT and SS are independent events, then from Eq. (6.6), we know that the proportion of persons in $NT \cap SS$ is given by

$$P(NT \cap SS) = P(NT)P(SS) = \left(\frac{350}{1000}\right)\left(\frac{100}{1000}\right)$$

$$= \frac{35,000}{1,000,000} = \frac{35}{1000}$$

Therefore, if 35/1000 of the persons involved in the experiment are expected to be in the event $NT \cap SS$, this means that 35/1000 of the 1000 persons, namely,

$$\frac{35}{1000} \times 1000 = 35$$

TABLE 6.5
Actual and expected contingency tables for allergy treatment data

	Actual Contingency Table				Expected Contingency Table		
	SS	MS	Row Σ		SS	MS	Row Σ
NT	44	306	350	NT	35	315	350
AC	19	131	150	AC	15	135	150
AS	37	463	500	AS	50	450	500
Column Σ	100	900	1000	Column Σ	100	900	1000

persons are expected to be in $NT \cap SS$. This argument accounts for the number 35 in the upper left corner of the expected contingency table of Table 6.5.

We can streamline the above calculation by noting in advance that the number of persons in $NT \cap SS$ will be $P(NT \cap SS) \times 1000$. Therefore, we can run the entire calculation directly as follows:

$$P(NT \cap SS) \times 1000 = P(NT)P(SS)(1000)$$

$$= \left(\frac{350}{1000}\right)\left(\frac{100}{1000}\right)(1000)$$

$$= \frac{350 \times 100}{1000} = 35$$

because of the cancellation of the last 1000 with one of those in the denominator.

We now proceed to calculate the expected number of persons in $AC \cap SS$, under the assumption of independence. We have

$$P(AC \cap SS) \times 1000 = P(AC)P(SS)(1000)$$

$$= \left(\frac{150}{1000}\right)\left(\frac{100}{1000}\right)(1000)$$

$$= \frac{150 \times 100}{1000} = 15$$

This calculation explains the appearance of the 15 in the expected contingency table of Table 6.5 at the intersection of the row headed "AC" and the column headed "SS."

The remaining cells of the expected contingency table are filled as follows:

$$P(AS \cap SS) \times 1000 = P(AS)P(SS)(1000)$$

$$= \left(\frac{500}{1000}\right)\left(\frac{100}{1000}\right)(1000)$$

$$= \frac{500 \times 100}{1000} = 50$$

$$P(NT \cap MS) \times 1000 = P(NT)P(MS)(1000)$$

$$= \frac{350 \times 900}{1000} = 315$$

$$P(AC \cap MS) \times 1000 = P(AC)P(MS)(1000)$$

$$= \frac{150 \times 900}{1000} = 135$$

$$P(AS \cap MS) \times 1000 = P(AS)P(MS)(1000)$$

$$= \frac{500 \times 900}{1000} = 450$$

For several reasons, including the ability to have a check on the calculation, it is useful to notice that the row and column totals in the expected contingency table are exactly the same as those in the actual contingency table. For example, in Table 6.5, the top rows in each of the contingency tables sum to 350, as

$$44 + 306 = 350$$

and

$$35 + 315 = 350$$

Because of the fact that the row and column sums must be the same in both contingency tables, it is possible to get away with computing only two of the six expected frequencies. This is, in fact, the way the "pros" (professional statisticians, of course) do it. In Table 6.6, for example, only the 35 and the 15 have been calculated using the formula for independent events. The remaining four cells have been filled by subtracting those two numbers from the row and column totals. For example,

$$350 - 35 = 315$$
$$150 - 15 = 135$$
$$100 - 35 + 15 = 50$$
$$500 - 50 = 450$$

The hand calculation of the components of the expected contingency table can be considerably simplified using this procedure. The number of cells for which the entire computation must be carried out is called the "degrees of freedom" of the contingency table. To find the degrees of freedom, we use the formula

★ $df = (r - 1)(c - 1)$ (6.8)

where r = number of rows in the table
c = number of columns in the table

The block of cells to be filled by computation consists of one less row and one less column than the original contingency table, and so contains $(r - 1)(c - 1)$ boxes. The contingency table for the allergy treatment data has three rows and two columns, so that $r = 3$ and $c = 2$. It follows that $df = (3 - 1)(2 - 1) = (2)(1) = 2$, so that it is necessary to directly fill only two[1] of the six boxes.

To decide the question of the dependence of the severity of the symptoms on the nature of the treatment, it remains only to measure the difference be-

[1]Not any two at all, but two chosen strategically!

TABLE 6.6
Expected frequency computations

	SS	MS	Row Σ
NT	35	315	350
AC	15	135	150
AS	50	450	500
Column Σ	100	900	1000

$$\frac{350 \times 100}{1000} = 35$$

$$\frac{150 \times 100}{1000} = 15$$

tween the actual and expected contingency tables of Table 6.5. We do this by means of the chi-square statistic. If we denote by f the number (actual frequency) appearing in a cell of the actual contingency table, and by e the number (expected frequency) appearing in the corresponding cell of the expected contingency table, then the chi-square statistic is given by the formula

$$\bigstar \quad \chi^2 = \Sigma \frac{(f - e)^2}{e} \qquad (6.9)$$

where there is one term in the sum for each cell in the table. If the classifications of treatment and severity of symptoms are truly independent, this fact would be indicated by each e being relatively close to its corresponding f. Therefore, in case of independence, the numerical value of χ^2 would be small. It follows that a large value of χ^2 would indicate a significant dependence relationship between treatment and severity of symptoms.

In testing

H: Treatment and symptoms are independent

versus

A: They are dependent

we would reject **H:** (independence) in favor of **A:** (dependence) at level α if $\chi^2 > \chi_\alpha^2 [df]$.

Using a significance level of $\alpha = .05$, and recalling that $df = (r - 1)(c - 1) = 2$ in this case, we see from Table A.5 that $\chi_{.05}^2[2] = 5.991$. (The relevant portion of Table A.5 can be found in Table 6.7.) Therefore we will reject **H** at level $\alpha = .05$ and conclude that the severity of symptoms depends on the treatment only if $\chi^2 > 5.991$. The calculations in Table 6.8 implement the formula for χ^2 above. The columns headed "f" and "e" are filled by the numbers appearing in the proper

TABLE 6.7
A small portion of Table A.5

df	$\chi_{.05}^2$	df
2	5.991	2

cells of the contingency tables of Table 6.5. From Table 6.8 it follows that

$$\chi^2 = \Sigma \frac{(f - e)^2}{e}$$

$$= \frac{(44 - 35)^2}{35} + \frac{(19 - 15)^2}{15} + \frac{(37 - 50)^2}{50} + \frac{(306 - 315)^2}{315}$$

$$+ \frac{(131 - 135)^2}{135} + \frac{(463 - 450)^2}{450}$$

$$= 2.3143 + 1.0667 + 3.3800 + .2571 + .1185 + .3756$$

$$= 7.5122$$

Therefore $\chi^2 = 7.512 > 5.991$, and so we reject **H:** (independence) at level $\alpha = .05$. We conclude that, according to the data, there seems to be a dependence relationship between the type of treatment and the severity of the symptoms. (Further medical research and analysis would be required to determine the nature of this dependence relationship.)

Example 6.5 Effectiveness of a New Drug Suppose the U-Feel Better (UFB) Pharmaceutical Company claims it has developed a drug which can ease mental depression. To check on the claim, the Food and Drug Administration (FDA) sets up a test involving 500 persons with mental depression. A random sample of 300 of them (the experimental group) are treated with the new drug, while the remaining 200 (the control group) are given only sugar pills. All 500 are given the

TABLE 6.8
Calculation of χ^2 for the allergy treatment data

Cell	f	e	f − e	(f − e)²	$\frac{(f - e)^2}{e}$
NT ∩ SS	44	35	9	81	2.3143
AC ∩ SS	19	15	4	16	1.0667
AS ∩ SS	37	50	−13	169	3.3800
NT ∩ MS	306	315	−9	81	.2571
AC ∩ MS	131	135	−4	16	.1185
AS ∩ MS	463	450	13	169	.3756
Sums	1000	1000	0		$\chi^2 = 7.5122$

TABLE 6.9
Contingency table of mental depression drug data

Treatment \ Reaction	Felt Better	No Effect	Felt Worse	Row Totals
New drug	50	240	10	300
Sugar pill	25	160	15	200
Column totals	75	400	25	500

impression that they are receiving the new treatment. All 500 persons are later asked whether the treatment they received made them feel better, made them feel worse, or had no effect. The resulting data are presented in Table 6.9.

SOLUTION   The statistician for the FDA considers $\alpha = .05$ to be a reasonable level of significance in testing

H: Independence (the drug is not effective)

versus

A: Dependence (the drug may be effective)

Here the hypothesis of independence means that a person's reaction does not depend on whether he was treated with the drug or with a sugar pill. If the treatment made no difference, we would be led to believe that the new drug was no more effective than a sugar pill and therefore not effective. Using the contingency table of Table 6.9, the statistician further observes that the degrees of freedom are

$$df = (r - 1)(c - 1) = (2 - 1)(3 - 1) = (1)(2) = 2$$

because there are two rows and three columns in the table. The chi-square test of independence, then, will have us reject **H** in favor of **A** at level $\alpha = .05$ if $\chi^2 > \chi_{.05}[2] = 5.991$. That number can be found in Table A.5 of the Appendix (and also in Table 6.7). We now proceed to the calculation of χ^2.
   The following events will be involved in the computational aspects of this problem:

ND = event that person was treated with new drug

SP = event that person was treated with sugar pill

FB = event that person felt better after treatment

NE = event that person had no effect from treatment

FW = event that person felt worse after treatment

TABLE 6.10
Actual and expected contingency tables of mental depression drug data

		Actual Contingency Table					Expected Contingency Table		
	FB	NE	FW	Row Σ		FB	NE	FW	Row Σ
ND	50	240	10	300	ND	45	240	15	300
SP	25	160	15	200	SP	30	160	10	200
Column Σ	75	400	25	500	Column Σ	75	400	25	500

We calculate the expected contingency table in the quickest way possible and present the results in Table 6.10. We have, under the assumption of independence, that

$$P(ND \cap FB) \times 500 = \frac{300 \times 75}{500} = 45$$

$$P(ND \cap NE) \times 500 = \frac{300 \times 400}{500} = 240$$

and we fill the remaining cells by subtracting these numbers from the row and column totals where called for. On the basis of the information presented in Table 6.10, we calculate the numerical value of χ^2 using the procedure outlined in Table 6.11.

Table 6.11 gives us the result that $\chi^2 = 5.556$, and it follows that $\chi^2 = 5.556 < 5.991$, and so we cannot reject the hypothesis of independence at level $\alpha = .05$. The results of the FDA, then, tend to refute the company's claim that their new drug relieves mental depression.

At this point in the discussion, the statistician for the pharmaceutical company challenges the conclusions of the FDA analysis. She suggests instead that all persons feeling no effect be eliminated from further consideration and only those persons indicating a change after the treatment be included in the compu-

TABLE 6.11
Calculation of χ^2 for the mental depression drug data

Cell	f	e	$f - e$	$(f - e)^2$	$\dfrac{(f - e)^2}{e}$
ND $\cap$ FB	50	45	5	25	.5556
SP $\cap$ FB	25	30	−5	25	.8333
ND $\cap$ NE	240	240	0	0	.0000
SP $\cap$ NE	160	160	0	0	.0000
ND $\cap$ FW	10	15	−5	25	1.6667
SP $\cap$ FW	15	10	5	25	2.5000
Sums	500	500	0		$\chi^2 = 5.5556$

TABLE 6.12
The company's view of mental depression drug data

Reaction / Treatment	Felt Better	Felt Worse	Row Totals
New drug	50	10	60
Sugar pill	25	15	40
Column totals	75	25	100

tations. Such a procedure is common, especially in advertising, where results of a survey are presented as being based on "all those responding to the question," or "all those expressing an opinion," or "all those indicating a preference," or some similar criterion for throwing away an often substantial portion of the data.

The company's statistician anchors her analysis on the contingency table illustrated in Table 6.12. This contingency table has $r = 2$ rows and $c = 2$ columns, so that the number of degrees of freedom is

$$df = (r - 1)(c - 1) = (2 - 1)(2 - 1) = 1$$

In the company view, then, the hypothesis **H:** (independence) should be rejected at level $\alpha = .05$ in favor of **A:** (dependence if χ^2 turns out to be larger than $\chi_{.05}^2[1] = 3.841$, as available in Table A.5 of the Appendix or the portion of it appearing in Table 6.13. In Table 6.14, we compare the actual and expected contingency tables based on the pharmaceutical company's view of the mental depression drug data.

The expected contingency table at the right side of Table 6.14 presents an especially clear illustration of the way in which such tables represent the independence of the two classifications, treatment and reaction to treatment. We observe first that 75 of the 100 persons under study (3 out of every 4) have reported that they felt better rather than worse after treatment. If reaction to treatment is truly independent of which treatment was received, this same proportion of persons feeling better (namely, 3 out of every 4) should be observed in each of the treatment groups. Therefore, of the 60 given the new drug, 45 should have reported feeling better, while of the 40 given sugar pills, 30 should have reported feeling better. The expected contingency table reflects these proportions exactly.

In Table 6.15, we calculate the numerical value of χ^2, which turns out to be

TABLE 6.13
A small portion of Table A.5

df	$\chi^2_{.05}$	df
1	3.841	1

TABLE 6.14

Actual and expected contingency tables of mental depression drug data based on the company's view of the statistics

	Actual Contingency Table				Expected Contingency Table		
	FB	FW	Row Σ		FB	FW	Row Σ
ND	50	10	60	ND	45	15	60
SP	25	15	40	SP	30	10	40
Column Σ	75	25	100	Column Σ	75	25	100

$x^2 = 5.556$. Because $x^2 = 5.556 > 3.841$, we are led to reject the hypothesis of independence at significance level $\alpha = .05$. Based on the company's view of the data, then, we would conclude that reaction does depend on treatment. The fact that more persons (50) actually felt better than the number of persons (45) that would be expected to feel better in case of independence tends to indicate that the new drug is somewhat effective in relieving mental depression.

What lessons can be drawn from comparing these two competing analyses of the mental depression drug data? When all 500 persons participating in the survey were included in the statistical analysis, the x^2 value of 5.556 did not permit us to reject the hypothesis of independence at level $\alpha = .05$, and so we had to conclude that there was insufficient evidence to substantiate the effectiveness of the new drug. However, upon elimination from further analysis of those 400 persons who reported that their treatments had no effect, the resulting x^2 value of 5.556 was sufficient to reject independence. What happened to cause these paradoxical results?

On the basis of purely practical considerations, the results of the mathematical analysis make good sense. If 400 of the 500 persons, including 240 of the 300 given the drug (a substantial majority of those participating in the survey), reported that the treatment they were given had no effect, then it cannot be considered likely that the new drug is really effective. Removing these 400 persons from consideration is equivalent to ignoring the fact that so many of those given the new drug were unaffected by it. After ignoring such a substantial percentage of those participating in the study, we should not be surprised at any

TABLE 6.15

The company's calculations of x^2 for the mental depression drug data

Cell	f	e	f − e	(f − e)²	$\dfrac{(f - e)^2}{e}$
ND ∩ FB	50	45	5	25	.5556
SP ∩ FB	25	30	−5	25	.8333
ND ∩ FW	10	15	−5	25	1.6667
SP ∩ FW	15	10	5	25	2.5000
Sums	100	100	0		$x^2 = 5.5556$

change of conclusion because whatever result happened to emerge would lack credibility. The procedure suggested by the company's statistician, then, is invalid because it excludes 80% of the data from consideration and bases its decision on a selected 20% of the data.

One final technical comment: It is a requirement of the proper usage of the chi-square test that there be a sufficiently large amount of data to guarantee that each expected frequency e be 5 or more. The specification that $e \geq 5$ is to be considered as an empirical rule of thumb, although it is supported by theoretical facts. Naturally, the more data there is available, the more reliable our statistical analysis will be. The requirement that $e \geq 5$ in every case merely puts some control on the number of data points considered acceptable.

EXERCISES 6.E

1   The following data records the positions of 400 members of the U.S. House of Representatives on repeal of a certain farm law:

	Position	
Party	For Repeal	Against Repeal
Republican	140	10
Democratic	100	150

At significance level $\alpha = .01$, do the data indicate that there is a dependence relationship between party affiliation and position on repeal of the law?

2   After a proposal was introduced in the city council to set up a car-free area in the downtown business district, a local newspaper surveyed 500 persons who regularly made use of that section of town, either for business or pleasure purposes. The objective of the survey was to find out whether persons who used their cars for transportation downtown felt differently about the proposal than those who ordinarily used public transportation. Each respondent was classified according to his or her usual method of transportation downtown and his or her opinion on the proposal for a car-free area. The results of the survey are as follows:

	Method of Transportation	
	Public Bus	Private Car
Favor proposal	200	100
Oppose proposal	100	100

At level $\alpha = .05$, can we conclude that a person's opinion of the proposal depends to some extent on his or her method of transportation?

3   An economist conducted a study of the effect of the presence of local water resources on local vegetable prices in 1000 localities west of the Mississippi

River. The resulting data relate a locality's availability of water with average retail vegetable prices in that locality.

Water Resources	Retail Vegetable Prices		
	Low	Moderate	High
Scarce	10	70	120
Satisfactory	40	400	60
Abundant	150	130	20

Are we justified in asserting at level $\alpha = .05$ that local vegetable prices depend to a significant extent on local water resources?

4   One psychologist has studied the incidence of grimacing and rheumatic disease in schizophrenic patients. In a study of 2000 such patients, he collected the following data:

Rheumatic Disease	Grimacing	
	Present	Not Present
Present	40	60
Not present	360	1540

At level $\alpha = .05$, does there seem to be a dependence between grimacing and rheumatic disease in schizophrenic patients?

5   A psychologist wants to know whether the fact of being male or female affects the ability to break the smoking habit. To answer the question, she selects a random sample of 70 men and 30 women from lists of participants in stop-smoking clinics and obtains the following data:

Smoking Status	Sex	
	Male	Female
Still not smoking	45	15
Returned to smoking	25	15

At significance level $\alpha = .01$, can the psychologist assert that the ability to stop smoking is independent of sex?

6   After an epidemic of a certain strain of flu, 1000 school children were examined for aftereffects. It turned out that 600 of the children had received medical treatment for the flu, while the other 400 had their cases unattended. The following data relates the occurrence of aftereffects with the receipt of medical treatment:

	Aftereffects	
	Noticeable	Not Noticeable
Was treated	225	375
Was not treated	175	225

At level $\alpha = .01$, can it be asserted that the treatment was of significant value in reducing aftereffects of the disease?

7 In a behavioral study of rats, each of 200 rats was classified by two characteristics, tail length and social behavior. The results appear in the following table:

Social Behavior	Tail Length		
	Short	Medium	Long
Aggressive	14	32	14
Normal	27	45	28
Submissive	9	23	8

Do the results of the study indicate at significance level $\alpha = .05$ that social behavior of rats depends on tail length?

8 A study recently conducted in the Los Angeles school system was aimed at determining whether or not an elementary pupil's ability to do mathematics was dependent on the teacher's ability to do algebra. One thousand students were selected at random from the system, and each was classified by his or her own ability to do elementary school mathematics and by the ability of his or her teacher to do algebra. The resulting data follows:

Teacher's Ability in Algebra	Pupil's Ability in Mathematics		
	Low	Average	High
Low	70	10	20
Average	260	330	110
High	70	60	70

Do the results of the study confirm at level $\alpha = .05$ that a pupil's ability to do mathematics depends on the teacher's ability to do algebra?

SUMMARY AND DISCUSSION

In Chap. 6, we returned to probability theory. Our direct concern has been the updating of probabilities of events, taking account of newly available information. We saw that the presence of new information represents a change in circumstances which can increase or decrease or have no effect on probabilities of particular events. If the new information does not affect the probability of an event in which we are interested, we say that our event is independent of the new information. We next studied contingency tables of cross-classified data for the purpose of trying to find out whether two descriptive classifications are independent of each other. Using the definition of independent events, we set up a companion expected contingency table of how the data would've turned out had the classification been truly independent. Then we used the chi-square test to measure the difference between the actual and the expected contingency tables. On the basis of the size of this difference, we made our decision as to whether or not the classifications could really be considered independent.

SUPPLEMENTARY EXERCISES

1 Decide whether each of the following statements is always true or sometimes false:
   (a) $P(A) + P(A^c) = 1$.
   (b) $P(A|B) + P(A^c|B) = 1$.
   (c) $P(A|B) + P(A|B^c) = 1$.
   (d) $P(A|B) + P(A^c|B^c) = 1$.
2 Anode Electronics, Betatron Atomics, and Cobalt Combinatorics are
   working feverishly but independently of each other to develop a fool-
   proof nuclear test detector. Their respective probabilities of success
   within five years are estimated as .4, .4, and .3. If those estimates are correct, what is the probability that at least one of them successfully de-
   velops the detector within 5 years.
3 You are offered the choice of one of the following two games to play:
   (a) You roll six fair dice, and you win if you roll at least one 6.
   (b) You roll twelve fair dice, and you win if you roll at least two 6's.
   Which game gives you the greater probability of winning?
4 You are offered the choice of one of the following two games to play:
   (a) You roll a fair die four times, and you win if you roll at least one 6.
   (b) You roll a pair of fair dice 24 times, and you win if you roll at least
        one 12.
   Which game gives you the greater probability of winning?
5 If we roll a fair tetrahedron (a four-sided, rather than a six-sided, die)
   once, the sample space for the experiment is $S = \{1, 2, 3, 4\}$, and all four
   outcomes are equally likely. Consider the events

   $A = \{1, 2\}$

   $B = \{1, 3\}$

   $C = \{1, 4\}$

   (a) Show that A, B, and C are pairwise independent, namely, that A is
        independent of B, B is independent of C, and C is independent of A.
   (b) Show, however, that A, B, and C do not constitute a set of indepen-
        dent events. Accomplish this by showing that $P(A \cap B \cap C)$ is not
        equal to $P(A)P(B)P(C)$.
   (c) Show that A is not independent of the event $B \cap C$.
6 If A and B are events such that $P(A) = 1/2$, $P(B) = 2/3$, and $P(A|B) = 1/4$,
   then calculate
   (a) $P(A^c|B)$
   (b) $P(A|B^c)$
   (c) $P(B|A)$
   (d) $P(A^c|B^c)$
7 A major corporation has several hundred projects in the initial stages of

development. These projects are classified into three groups: group *S* for those having a 90% chance of eventually turning a profit, group *M* for those having a 70% chance of turning a profit, and group *R* for those having a 50% chance of turning a profit. Of the corporation's projects, 30% are in group *S*, 50% are in group *M*, and 20% are in group *R*. What percentage of the corporation's projects will eventually turn a profit?

8 Three years out of every 10, one southern state suffers both coastal hurricanes and inland flooding. Four years out of every 10, it suffers coastal hurricanes but no inland flooding.
   (a) What percentage of years does the area suffer coastal hurricanes?
   (b) For this year, the weather service has forecasted coastal hurricanes. What is the probability that inland flooding will also occur?

9 In a recent election, the local Democratic cadidate received 50% of the vote, the Republican candidate got 40%, while an independent candidate obtained 10%. At the same time, ballot proposition 23 received the votes of 20% of those who voted Democratic, 90% of those who voted Republican, and 90% of those who voted independent. What proportion of the voters favored proposition 23?

10 A stop-smoking clinic advertises that, in a test region last August, 80% of those who tried and were able to stop smoking had participated in its program, while only 30% of those who tried and failed to stop smoking had participated. Further research reveals that, of all smokers in the region who tried to kick the habit last August, 90% had failed, while only 10% succeeded. What proportion of those who had attempted to stop smoking were customers of the clinic?

11 A psychological test is designed to separate entering college freshmen into good prospects and not-so-good prospects. Among those who later performed satisfactorily during the year, 80% had passed the test. Among the students who did unsatisfactory work in their first year, only 40% had passed the test. On the whole, 70% of the students tested did satisfactory work that year. What percentage of entering freshmen passed the psychological test?

12 The noxious oxides of nitrogen comprise 20% of all pollutants in the air, by weight, in a certain metropolitan area. Automobile exhaust accounts for 70% of those noxious oxides but only 10% of all other pollutants in the air. Of all the pollution in the air, what percentage is contributed by automobile exhaust?

13 By his own admission, an unusually candid student knew only 50% of the material covered in his geography class. The final exam was a multiple-choice test, having four possibilities for the answer to each question, and it covered in great detail all aspects of the course. On the half of the test covering material the student knew, he naturally got all the answers right; on the other half, he guessed one of the four options. What proportion of the questions did the student answer correctly?

14 In a recent small war, one side used both heat-seeking missiles and laser-guided rockets to shoot down enemy aircraft. In fact, the objects launched at enemy planes were 80% of the heat-seeking type and 20% of the laser-guided variety. The heat-seeking missiles struck their targets 40% of the time, while the laser-guided rockets scored hits 90% of the time. What

percentage of the objects launched at enemy aircraft actually struck their targets?

15 An independent supermarket receives 60% of its fruit from Foster's Fruit Fields and the other 40% from Gregory's Giant Groves. On the average, about 25% of Foster's shipments arrive overripe, while only 16% of Gregory's do. Because the farmers issue credit when their fruit arrives overripe, the market keeps strict records on the subject. What percentage of the market's fruit shipments arrive overripe?

16 A test for a certain rare disease is capable of detecting the disease in 97% of all afflicted individuals. However, when entirely healthy persons are tested, 5% of them are incorrectly diagnosed as having the disease. When persons who have other milder diseases are tested, 10% of them are incorrectly diagnosed as having the rare disease. In reality, only about 1% of the population has the rare disease, 3% have other milder diseases, and the remaining 96% are entirely healthy. What proportion of the population is diagnosed by the test as having the rare disease?

. 17 One financial analyst believes that the stock market's Dow-Jones Average (DJA) declines the week following an increase in the prime rate of interest at major banks. He feels, on the other hand, that the DJA rises the week after the prime rate goes down. To substantiate his opinion, he studies the data for the last 1000 weeks and obtains the following information relating the prime rate behavior one week with the DJA behavior the next:

	Dow-Jones Average a Week Later		
Prime Rate	Down	Same	Up
Down	100	50	250
Same	70	100	30
Up	330	50	20

At level $\alpha = .01$, can the financial analyst conclude that changes in the DJA really depend on fluctuations in the prime rate of interest?

18 In a recent study of the use of corporate annual financial reports by prospective and current stockholders, it was hypothesized that college-educated investors were better able to understand a company's balance sheet than were those investors who had not had a college education. To check on the hypothesis, 400 randomly selected stockholders were interviewed and classified according to whether or not they had a college education and whether or not they had difficulty understanding the balance sheet. The results of the survey follow:

	Had Difficulty in Understanding Company Balance Sheet	
	Yes	No
Had college education	145	55
Had no college education	155	45

At level $\alpha = .05$, do the results of the survey indicate that college-educated investors differ from other investors in ability to understand the balance sheet?

19 A dental study of 200 children, aimed at analyzing the effect of fluoride toothpaste on the number of cavities, yielded the following data on the change in the number of cavities each child had, compared with the total at the last checkup:

	More Cavities	Same Amount	Fewer Cavities	
Fluoride toothpaste	5	5	40	50
Regular toothpaste	20	20	60	100
No toothpaste	20	20	10	50
	45	45	110	

Decide at level $\alpha = .05$ whether or not the study shows that the change in number of cavities depends significantly on the type of toothpaste used.

7

THE ANALYSIS
OF VARIANCE

In our discussion of testing statistical hypotheses in Chap. 4, we studied one-sample tests and two-sample tests about population means. We introduced some techniques valid for large samples and some for small samples, and we allowed for independent samples and for paired samples. However, as you may have noticed, we did not analyze a situation involving three or four or more samples. Yet, it is often necessary to run an experiment involving several samples. For example, an agricultural seed company interested in developing multipurpose seeds might want to know whether a new variety of corn will produce similar or differing yields in widely scattered geographical regions, such as low plains, delta, mountains, and irrigated desert. The way to find out would be to plant several acres in each of the four regions and then compare the four mean yields. As a second example, an office executive might be considering five competing computerized storage systems for his company's office records, and he would want to know if they are equally efficient or not. The problem could be analyzed by solving some sample information retrieval problems in each of the five systems and then comparing the five mean efficiency levels. As it turns out, none of the techniques studied in Chap. 4 are effective in studying either of these and similarly structured problems. We need a completely new method of analysis, a method of testing the equality of several means by a several-sample test. That method, "the analysis of variance," is our subject for this chapter.

SECTION 7.A
COMPONENTS OF THE VARIANCE

If we plant corn on five randomly selected acres in each of four types of geographical regions, we will have a total of twenty plots of corn involved in our experiment. Even if there is no real difference between the corn yields in differing geographical regions, it would be unreasonable to expect all 20 plots to give exactly the same yield. As we have discussed earlier in Chap. 5, for example, this would be as unreasonable as expecting an entire set of data points to lie exactly on a straight line, even if the underlying relationship were truly linear. Therefore we would expect to have some variation among the corn yields, and it is the extent of this variation that leads us to decide whether or not there is a significant difference in yield among different types of geographical regions.

We denote the true mean yields of the new variety of corn in the various geographical regions as follows:

μ_L = true mean yield in low plains

μ_D = true mean yield in delta

μ_M = true mean yield in mountains

μ_I = true mean yield in irrigated desert

We would then want to test at significance level α the hypothesis

H: $\mu_L = \mu_D = \mu_M = \mu_I$

against the alternative

A: These true means are not all the same

Suppose our experiment results in the data presented in Table 7.1. How are we to decide whether or not the true mean yields are the same for the four geographical regions? As can be seen from the data of Table 7.1, we have

$\bar{x}_L$ = sample mean yield in low plains = 8

$\bar{x}_D$ = sample mean yield in delta = 8

$\bar{x}_M$ = sample mean yield in mountains = 6

$\bar{x}_I$ = sample mean yield in irrigated desert = 5

TABLE 7.1
Corn yields in different geographical regions
(Data given in hundreds of bushels)

Low Plains	Delta	Mountains	Irrigated Desert	Entire Set of Data
13	8	3	4	
6	2	5	8	
5	10	10	5	
6	9	6	6	
10	11	6	2	
Sums 40	40	30	25	135
Sample means 8	8	6	5	6.75

We notice immediately that these four sample means are not all the same. Before we draw any inferences regarding the hypothesis that the true means are identical, we have to discuss two pertinent questions.

1 To what extent do the four sample means differ among themselves?
2 To what extent do the four sample means represent the true means of their respective groups?

We are looking for conditions under which we will be justified in rejecting the hypothesis

H: $\mu_L = \mu_D = \mu_M = \mu_I$

in favor of the alternative

A: These true means are not all the same

Our decision to accept or reject **H** will be made on the basis of the answers to the above two questions. We will be inclined to reject **H** if the sample means differ greatly among themselves and we think that they are accurate representations of the true means; otherwise, we will probably accept **H**. In Table 7.2, we illustrate the relationship between the answers to questions 1 and 2 above and our decision regarding the hypothesis **H**.

All that remains, therefore, is to analyze the data of Table 7.1 with a view toward answering the two questions above.

Let's deal first with the first question involving the extent to which the four sample means differ among themselves. We have the sample means 8, 8, 6, and 5, each based on one of the four sets of data in Table 7.1. The extent to which these sample means differ among themselves can be measured by their variation from *their own* mean 6.75, which is also the "grand mean" of all 20 data

TABLE 7.2

When should the hypothesis H: $\mu_L = \mu_D = \mu_M = \mu_I$ **be rejected in favor of A: These true means are not all the same?**

Extent to Which Sample Means Represent True Means	Extent to Which Sample Means Differ among Themselves	
	Little	Much
Little	Cannot reject **H**	Cannot reject **H**
Much	Cannot reject **H**	Should reject **H**

points. We can denote this measure of variation by

$$VB = (8 - 6.75)^2 + (8 - 6.75)^2 + (6 - 6.75)^2 + (5 - 6.75)^2$$

$$= (1.25)^2 + (1.25)^2 + (-.75)^2 + (-1.75)^2$$

$$= 1.5625 + 1.5625 + .5625 + 3.0625 = 6.75$$

where VB stands for "variation between the samples."

Before we investigate the significance of VB in detail, let's work with the second question, which concerns the extent to which the four sample means represent the true means of their groups. To study this question, we can compute the variation within each sample from *its own* mean as follows (using the squared deviations from the mean as we have been doing since Sec. 1.D):

$$VW_L = (13 - 8)^2 + (6 - 8)^2 + (5 - 8)^2 + (6 - 8)^2 + (10 - 8)^2$$

$$= 25 + 4 + 9 + 4 + 4 = 46$$

$$VW_D = (8 - 8)^2 + (2 - 8)^2 + (10 - 8)^2 + (9 - 8)^2 + (11 - 8)^2$$

$$= 0 + 36 + 4 + 1 + 9 = 50$$

$$VW_M = (3 - 6)^2 + (5 - 6)^2 + (10 - 6)^2 + (6 - 6)^2 + (6 - 6)^2$$

$$= 9 + 1 + 16 + 0 + 0 = 26$$

$$VW_I = (4 - 5)^2 + (8 - 5)^2 + (5 - 5)^2 + (6 - 5)^2 + (2 - 5)^2$$

$$= 1 + 9 + 0 + 1 + 9 = 20$$

It follows that the "variation within the samples," denoted by VW, can be calculated as

$$VW = VW_L + VW_D + VW_M + VW_I$$

$$= 46 + 50 + 26 + 20 = 142$$

Before proceeding, let's take a look at what we have done so far. We have four samples of data, and we have defined two measures of variation.

VB = variation between the samples

VW = variation within the samples

Now if the hypothesis is true, there should be little variation *between* the samples, but if **H** is false, there should be a lot of variation *between* the samples. What does this have to do with variation *within* the samples? Well, as it turns out, there are only two types of variation—variation between samples and variation within samples. Therefore when VB is small, VW is large, and when VB is large, VW is small. *When the hypothesis* **H** *is ture, VB is small and VW is large, and when the alternative* **A** *is true, VB is large and VW is small.* This is the basic principle upon which the method of analysis of variance is based. To see how it works on the data of Table 7.1, it is necessary to compute the "total variation," denoted by TV, of the data and then to compare it with VB and VW. In Table 7.3, we compute the total variation of the data, and we come up with the result that $TV = 175.75$.

If we denote by n the number of data points in each sample, then technical algebraic calculations show that the relationship between TV, VB, and VW is given by the equation $TV = nVB + VW$.

For the data of Table 7.1, we have shown that

$$TV = 175.75$$

$$n = 5$$

$$VB = 6.75$$

$$VW = 142$$

The equation $TV = nVB + VW$ then reduces to the assertion that

$$175.75 = (5)(6.75) + 142$$

which is a true statement since $(5)(6.75) = 33.75$. The formula

$$\bigstar \quad TV = nVB + VW \tag{7.1}$$

is called the "components of the variance formula." In the next section we shall see how a revised version of it can be used to simplify and organize the analysis of variance calculations. In the remainder of this section, however, we shall simply complete the discussion of the corn yield problem by formally testing the hypothesis of interest.

To test **H** against **A**, we use the F test, the rejection rules for which are based

TABLE 7.3
Calculation of the total variation
(Corn yield data)

x	$x - \bar{x}$	$(x - \bar{x})^2$
13	6.25	39.0625
6	−.75	.5625
5	−1.75	3.0625
6	−.75	.5625
10	3.25	10.5625
8	1.25	1.5625
2	−4.75	22.5625
10	3.25	10.5625
9	2.25	5.0625
11	4.25	18.0625
3	−3.75	14.0625
5	−1.75	3.0625
10	3.25	10.5625
6	−.75	.5625
6	−.75	.5625
4	−2.75	7.5625
8	1.25	1.5625
5	−1.75	3.0625
6	−.75	.5625
2	−4.75	22.5625
135	0	175.7500

$\bar{x} = \frac{135}{20} = 6.75$

$TV = 175.75$

on the table of the F distribution appearing in Table A.6 of the Appendix. The rejection rule is as follows: We reject **H** in favor of **A** at level α if

$$F > F_\alpha [m - 1, m(n - 1)]$$

where

$$\bigstar \quad F = \frac{nVB/(m - 1)}{VW/m(n - 1)} \tag{7.2}$$

In the expression for F, we have used the symbols

m = number of samples involved in the problem

n = number of data points in each sample

DEGREES OF FREEDOM OF THE F DISTRIBUTION

A word about the F distribution is now in order. The percentage points of the F distribution are of the form

$$F_\alpha[\text{dfn, dfd}]$$

where dfn = degrees of freedom in numerator

dfd = degrees of freedom in denominator

The fact that the percentage points of the F distribution are classified by two indices of degrees of freedom contrasts markedly with those of the t and chi-square distributions which are classified by only one index of degrees of freedom. In analysis of variance, particularly the corn yield example,

$$\text{dfn} = m - 1 = 4 - 1 = 3$$

because 4 samples were used in the problem. Also

$$\text{dfd} = m(n - 1) = 4(5 - 1) = 16$$

because there are 5 data points per sample. Therefore if we use a level of significance $\alpha = .05$, then we agree to reject **H** if

$$F > F_{.05}\ [3,\ 16] = 3.24$$

because $m = 4$ and $n = 5$ here. In Table 7.4, we exhibit a small portion of Table A.6, showing where to find $F_{.05}[3, 16]$. Now, recalling that

$$nVB = (5)(6.75) = 33.75$$

$$VW = 142$$

$$m - 1 = 4 - 1 = 3$$

$$m(n - 1) = 4(5 - 1) = 16$$

we have

$$F = \frac{nVB/(m - 1)}{VW/m(n - 1)} = \frac{33.75/3}{142/16} = \frac{(33.75)(16)}{(142)(3)} = 1.27$$

We agreed to reject **H** if $F > 3.24$. But actually it turned out that $F = 1.27 < 3.24$, and so we cannot reject **H** at level $\alpha = .05$. We therefore conclude that the data

TABLE 7.4
A small portion of Table A.6

	Degrees of Freedom for Numerator	
Degrees of freedom		3
for denominator	16	3.24

supports the hypothesis that the mean corn yields do not differ significantly among the four geographical regions.

How were the components of the variance involved in the solution to the above problem? Well, the final decision was that F was not large enough to permit us to reject **H**. Looking at the formula for F, we can see that this means that VB was not large enough relative to VW to warrant the rejection of **H** in favor of **A**. In other words, the variation between the sample means was not much greater than the variation within the samples, and therefore we could not conclude that variations among the data points could be attributed largely to variations between the sample means. We were therefore led to believe that the sample means did not exhibit inordinately large differences among themselves.

EXERCISES 7.A

1 An individual considering purchase of a 5000-dollar life insurance policy wants to find out if there are substantial differences between insurance premiums of the various types of policies available. In particular, he investigates the prices for the following related types of policies: individual renewable term, individual convertible term, group renewable term, and government group insurance. For each of these four types of policies, he obtains price quotations from six different competing underwriters. These price quotations are as follows:

Price Quotations for Monthly Premium, dollars			
Individual Renewable Term	Individual Convertible Term	Group Renewable Term	Government Group Insurance
2.50	2.00	2.00	1.00
3.00	2.20	1.50	1.20
2.20	1.80	1.80	1.00
2.80	2.50	2.00	1.10
2.50	3.00	2.00	1.50
3.00	2.00	1.80	1.40

(a) Calculate the components of the variance, and write down the numbers appearing in the components of the variance formula $TV = nVB + VW$.

(b) Decide at level of significance $\alpha = .05$ whether or not there are significant differences between the average monthly premiums of the four types of policies under study.

2 A stock brokerage firm, member of the New York Stock Exchange and other major exchanges, thinks it has a psychological test which can pick out those individuals who will turn out to be successful stockbrokers. As the firm usually recruits its employees from outside the securities industry, it would like to know in which occupations it can expect to find those with a high level of aptitude, as determined by the psychological test. Before the current recruiting drive begins, the firm therefore decides to give its psychological test to six randomly selected individuals in each of six occupational classifications. The aptitude scores follows:

Aptitude Scores of Persons in Six Occupational Classifications					
Commission Salesperson	Office Clerk	Teacher	Civil Servant	Career Military	Professional Athlete
80	70	80	40	50	70
70	80	60	90	10	60
80	60	90	50	50	30
90	30	70	40	60	50
60	50	80	30	30	80
40	70	10	20	40	70

(a) Calculate the components of the variance.
(b) At level $\alpha = .01$, can the stock brokerage firm conclude that there is a significant difference in aptitude among persons in the six occupational classifications?

3 An ichthyologist interested in the development of coastal nuclear-generating facilities wants to find out if temperature changes in the ocean's water will have a significant effect on the growth of fish indigenous to the region. He sets up an experiment involving four groups of seven recently hatched specimens each of the same species of fish. Each group is placed in a simulated ocean environment in which all factors are controlled and identical, with the exception of the temperature of the water. Six months later, each of the 28 specimens is weighed, and their weights are recorded in the following data:

Weights of Specimens at Four Water Temperatures, ounces			
40°F	42°F	44°F	46°F
20	18	24	16
18	25	17	17
23	16	14	26
16	20	22	19
15	24	25	14
25	16	27	20
16	21	18	21

(a) Calculate the components of the variance.
(b) At level $\alpha = .05$, do the results of the experiment indicate that the temperature differences tested have a significant effect on the average weight of fish in the region?

4 As part of a study of the manner in which pork-barrel projects are funded, a political scientist gathered the following information from randomly selected federal budgets over the past 40 years. The data points are the total appropriation in the budgets for pork-barrel projects in each of the three socio-politico-economic subdivisions of the nation: urban, suburban, and rural.

Appropriation for Each Socio-Politico-Economic Subdivision, millions of dollars		
Urban	Suburban	Rural
8	4	11
10	8	10
12	9	8
15	10	9
18	12	10
24	17	12
30	25	16
27	35	20

(a) Calculate the components of the variance.
(b) At level $\alpha = .01$, can we conclude from the data that the average amounts of federal money dispensed to each of the subdivisions are approximately the same?

SECTION 7.B
TESTING FOR DIFFERENCES AMONG SEVERAL MEANS

We now proceed to the development of the techniques of "one-way analysis of variance," a computational method of testing for differences among the true means of several groups. The philosophy underlying one-way analysis of variance has already been dealt with in the previous section. Here we will concentrate on constructing a more efficient way of organizing the data and computations needed in carrying out the F test for the equality of several means.

Let's begin by taking another look at the corn yield example of Sec. 7.A. In order to test the hypothesis

$$\text{H:} \quad \mu_L = \mu_D = \mu_M = \mu_I$$

against the alternative

A: These true means are not all the same

it was necessary to compute the number

$$F = \frac{nVB/(m - 1)}{VW/m(n - 1)}$$

From the data of Table 7.1, it was easy to see that $n = 5$ and $m = 4$, and it was not really difficult to calculate $VB = 6.75$. The calculation of VW, however, was another story. That required us to do five separate calculations, one each of VW_L, VW_D, VW_M, and VW_I, and one summing all the latter quantities. You must admit that the work required to calculate VW was somewhat out of line with that needed for the remainder of the problem, and that a shortcut method of computing VW would be much appreciated by all those working in the field (even by those having access to expensive calculators). The level of general appreciation would increase as the number of samples involved grow, for the work required to compute VW by the method of Sec. 7.A would grow correspondingly.

As it turns out, there is a very simple way of getting the value of VW. It is based on the components of the variance formula studied in the previous section, which asserts that

$$TV = nVB + VW$$

The shortcut formula is an immediate algebraic consequence of the above, namely, the assertion that

$$VW = TV - nVB$$

As we have already pointed out, n and VB are relatively easy to compute, and so if we can come up with a quick way of getting TV, we will have done the job.

Fortunately, there is a quick way of calculating TV which is well suited to both machine and hand computation. The explicit formula for TV is similar to the shortcut formula (3.3) for the standard deviation that we first discussed way back in Chap. 1 and later made extensive use of in Chaps. 3 and 4. In particular, TV is given by the formula

$$TV = \Sigma x^2 - \frac{(\Sigma x)^2}{N}$$

where N = total number of data points in all samples

Σx^2 = sum of the squares of all N data points

Σx = sum of all N data points

Example 7.1 Word Processing Systems   In Table 7.5, we have listed the weekly costs of operating each of five competing word processing (WP) systems in offices of a major steel corporation. The data were taken on randomly selected

TABLE 7.5
Weekly costs of different *WP* systems
(Data given in hundreds of dollars)

	United Electronics	Verbiage Control	Western Words	X-Pensive Palaver	Yorktown Machine	Entire Set of Data
	12	12	3	14	16	57
	13	10	6	18	21	68
	7	6	10	18	15	56
	8	4	13	14	20	59
Sums	40	32	32	64	72	240

$\{ \bar{X} = 60$

$\bar{X}$ 10   8   8   16   18

weeks during a 6-month trial period. Costs are only one factor influencing the
corporation's choice of a *WP* system, and so the corporation would like to know
whether or not costs run about the same for each of the five systems.

SOLUTION We use the following symbols to denote the true means involved in
this example:

  μ_U = true mean cost of United Electronics system

  μ_V = true mean cost of Verbiage Control system

  μ_W = true mean cost of Western Words system

  μ_X = true mean cost of X-Pensive Palaver system

  μ_Y = true mean cost of Yorktown Machine system

  We want to test the hypothesis

  H: $\mu_U = \mu_V = \mu_W = \mu_X = \mu_Y$

against the alternative

  A: These true means are not all the same

Recalling that

  m = number of samples involved in the problem

  n = number of data points in each sample

we see that

  $m = 5$

  $n = 4$

TABLE 7.6
A small portion of Table A.6

	Degrees of Freedom for Numerator	
		4
Degrees of freedom for denominator	15	3.06

Since the general rule is to reject **H** in favor of **A** at level α if

$$F > F_\alpha[m - 1, m(n - 1)]$$

we should reject **H** at level $\alpha = .05$ if

$$F > F_{.05}[4, 15] = 3.06$$

(We present in Table 7.6 the relevant portion of Table A.6.) Therefore it remains only to compute

$$F = \frac{nVB/(m - 1)}{VW/m(n - 1)}$$

In Table 7.7, we set up the data of Table 7.5 for the computation of *TV*. (A somewhat related construction in Table 7.3 gave a slightly longer calculation of *TV* for a different set of data.) The computations in Table 7.7 show that

$$N = 20$$

$$\Sigma x = 240$$

$$\Sigma x^2 = 3398$$

for the *WP* systems data, from which it follows that

$$TV = \Sigma x^2 - \frac{(\Sigma x)^2}{N} = 3398 - \frac{(240)^2}{20}$$

$$= 3398 - \frac{57,600}{20} = 3398 - 2880 = 518$$

As long as we are on the subject of shortcut formulas, let's complete the story by presenting and showing how to use a shortcut formula for computing the term *nVB*. Our calculational task will be immediately shortened by one step simply because we will be getting the term *nVB* in its entirety rather than merely *VB* alone. The shortcut formula for *nVB* is

$$nVB = \Sigma\left(\frac{S^2}{n}\right) - \frac{(\Sigma x)^2}{N}$$

where S is the sum of the n data points of a sample. Referring to Table 7.5, we can see that

$$S_U = 40$$

$$S_V = 32$$

$$S_W = 32$$

$$S_X = 64$$

$$S_Y = 72$$

Because $\Sigma x = 240$, $N = 20$, and $n = 4$ as before, the shortcut formula for nVB

TABLE 7.7
Calculation of the total variation TV
(*WP* systems data)

x	x^2
12	144
13	169
7	49
8	64
12	144
10	100
6	36
4	16
3	9
6	36
10	100
13	169
14	196
18	324
18	324
14	196
16	256
21	441
15	225
20	400
Sums 240	3398

$N = 20$
$\Sigma x = 240$
$\Sigma x^2 = 3398$

$$TV = \Sigma x^2 - \frac{(\Sigma x)^2}{N} = 3398 - \frac{(240)^2}{20} = 518$$

yields

$$nVB = \left(\frac{40^2}{4} + \frac{32^2}{4} + \frac{32^2}{4} + \frac{64^2}{4} + \frac{72^2}{4}\right) - \frac{(240)^2}{20}$$

$$= \left(\frac{1600}{4} + \frac{1024}{4} + \frac{1024}{4} + \frac{4096}{4} + \frac{5184}{4}\right) - \frac{57{,}600}{20}$$

$$= (400 + 256 + 256 + 1024 + 1296) - 2880$$

$$= 3232 - 2880$$

$$= 352$$

As we have just observed, $TV = 518$ and $nVB = 352$. It immediately follows that

$$VW = TV - nVB = 518 - 352 = 166$$

Therefore

$$F = \frac{nVB/(m-1)}{VW/m(n-1)} = \frac{352/4}{166/15} = \frac{88}{11.1} = 7.93$$

Because we have agreed to reject **H** in favor of **A** at level of significance $\alpha = .05$ if F turned out to be greater than 3.06, the result that $F = 7.93 > 3.06$ leads us to reject **H**. Therefore, at level $\alpha = .05$, we conclude that the costs of the five WP systems are not all the same.

In Table 7.8 we illustrate how to combine the calculations made in Tables 7.5 and 7.7 into a single table and thereby to obtain a more direct computation of TV, nVB, and ultimately VW.

Example 7.2 Gasoline Efficiency   A motorist would like to find out whether or not he gets the same gasoline efficiency (in miles per gallon) regardless of the brand of fuel he uses. He chooses five brands of gasoline available in his locality, and he records the number of miles per gallon he obtained on several tankfuls of each brand. In Table 7.9, his recorded mileage (miles per gallon) for each tankful is listed.

SOLUTION   The items of primary interest are the five true means

μ_T = true mean number of miles per gallon of Thrifty gasoline

μ_B = true mean number of miles per gallon of Broadway gasoline

μ_F = true mean number of miles per gallon of Federated gasoline

μ_G = true mean number of miles per gallon of Gibraltar gasoline

μ_H = true mean number of miles per gallon of Holiday gasoline

TABLE 7.8
One-way analysis of variance calculations
(*WP* systems data)

x^2	x	S	S^2	n	S^2/n
144	12				
169	13	40	1600	4	400
49	7				
64	8				
144	12				
100	10	32	1024	4	256
36	6				
16	4				
9	3				
36	6	32	1024	4	256
100	10				
169	13				
196	14				
324	18	64	4096	4	1024
324	18				
196	14				
256	16				
441	21	72	5184	4	1296
225	15				
400	20				
3398	240			20	3232

$$TV = \Sigma x^2 - \frac{(\Sigma x)^2}{N} = 3398 - \frac{(240)^2}{20} = 518$$

$$nVB = \Sigma\left(\frac{S^2}{n}\right) - \frac{(\Sigma x)^2}{N} = 3232 - \frac{(240)^2}{20} = 352$$

$$VW = TV - nVB = 518 - 352 = 166$$

TABLE 7.9
Efficiency of different brands of gasoline
(Data in miles per gallon)

Thrifty	Broadway	Federated	Gibraltar	Holiday
26	29	29	31	28
25	27	29	28	23
28	26	26	26	24
24	30	24	28	28
29	25	25	26	25
24	25	23	27	22
26		26	28	25
			30	
		n		
7	6	7	8	7

We want to test the hypothesis

$$H: \quad \mu_T = \mu_B = \mu_F = \mu_G = \mu_H$$

against the alternative

A: These true means are not all the same

In one-way analysis of variance, we compute[1]

$$\bigstar \quad F = \frac{nVB/(m-1)}{VW/(N-m)} \tag{7.3}$$

and we reject **H** in favor of **A** at level α if

$$\bigstar \quad F > F_\alpha[m-1, N-m] \tag{7.4}$$

In this gasoline efficiency example

$$m = \text{number of samples} = 5$$

and

$$N = \text{total number of data points}$$
$$= n_T + n_B + n_F + n_G + n_H$$
$$= 7 + 6 + 7 + 8 + 7 = 35$$

Therefore the degrees of freedom for the appropriate percentage point of the F distribution are

$$m - 1 = 5 - 1 = 4 \qquad \text{for the numerator}$$

and

$$N - m = 35 - 5 = 30 \qquad \text{for the denominator}$$

If we consider a level of significance of $\alpha = .01$ to be appropriate for this problem, then we agree to reject **H** in favor of **A** if

$$F > F_{.01}[4, 30] = 4.02$$

It therefore remains only to calculate the numerical value of F and to observe whether or not it exceeds 4.02. Inserting the numbers $m = 5$ and $N = 35$, we

---

[1] If all n's are equal, then $N = mn$, so that $N - m = mn - m = m(n-1)$, the number which appeared in formula (7.2).

have

$$F = \frac{nVB/(m-1)}{VW/(N-m)} = \frac{nVB/4}{VW/30} = \frac{30nVB}{4VW}$$

where

$$nVB = \sum\left(\frac{S^2}{n}\right) - \frac{(\Sigma x)^2}{N}$$

$$TV = \Sigma x^2 - \frac{(\Sigma x)^2}{N}$$

so that

$$VW = TV - nVB$$

$$= \left[\Sigma x^2 - \frac{(\Sigma x)^2}{N}\right] - \left[\sum\left(\frac{S^2}{n}\right) - \frac{(\Sigma x)^2}{N}\right]$$

$$= \Sigma x^2 - \sum\left(\frac{S^2}{n}\right)$$

In Table 7.10, we present the calculations required to obtain the value of F, especially the calculation of

$$\bigstar \quad nVB = \sum\left(\frac{S^2}{n}\right) - \frac{(\Sigma x)^2}{N} \tag{7.5}$$

and

$$\bigstar \quad VW = \Sigma x^2 - \sum\left(\frac{S^2}{n}\right) \tag{7.6}$$

We see from the results of Table 7.10 that

$$nVB = 38.57$$

$$VW = 130$$

from which it follows that

$$F = \frac{30nVB}{4VW} = \frac{(30)(38.57)}{(4)(130)} = 2.23$$

As we have agreed to reject the hypothesis

$$\textbf{H:} \quad \mu_T = \mu_B = \mu_F = \mu_G = \mu_H$$

TABLE 7.10
One-way analysis of variance calculations
(Gasoline efficiency data)

Sums of x^2	x^2	x	S	S^2	n	S^2/n
	676	26				
	625	25				
	784	28				
	576	24	182	33,124	7	4732
	841	29				
	576	24				
4754	676	26				
	841	29				
	729	27				
	676	26	162	26,244	6	4374
	900	30				
	625	25				
4396	625	25				
	841	29				
	841	29				
	676	26				
	576	24	182	33,124	7	4732
	625	25				
	529	23				
4764	676	26				
	961	31				
	784	28				
	676	26				
	784	28	224	50,176	8	6272
	676	26				
	729	27				
	784	28				
6294	900	30				
	784	28				
	529	23				
	576	24				
	784	28	175	30,625	7	4375
	625	25				
	484	22				
4407	625	25				
Sums	24,615	925			35	24,485

$$nVB = \Sigma\left(\frac{S^2}{n}\right) - \frac{(\Sigma x)^2}{N} = 24,485 - \frac{(925)^2}{35} = 38.57$$

$$VW = \Sigma x^2 - \Sigma\left(\frac{S^2}{n}\right) = 24,615 - 24,485 = 130$$

in favor of the alternative

A: These true means are not all the same

at level $\alpha = .01$, if F turned out to be larger than 4.02, the result that

$$F = 2.23 < 4.02$$

indicates that we should not reject **H**. At significance level $\alpha = .01$, then, we conclude that the data do not reveal significant differences in mileage per gallon between the five brands of gasoline tested.

Before we leave the subject of analysis of variance, it would be useful to explain what we have done in terms of the traditional language and vocabulary. Referring again to the basic gasoline efficiency data of Table 7.9, we can view the five different brands of gasoline as "treatments" affecting the car's mileage per gallon. Under analysis of variance restrictions analogous to those of the t test discussed at the end of Sec. 4.C, each treatment would yield a set of data whose underlying population is normally distributed, and all such populations would have the same true standard deviation σ. The hypothesis

$$\textbf{H:} \quad \mu_T = \mu_B = \mu_F = \mu_G = \mu_H$$

asserts that all such populations also have the same mean μ (which is, of course, the common numerical value of μ_T, μ_B, μ_F, μ_G, and μ_H), and therefore that *all* the data points come from a *single* normally distributed population having mean μ and standard deviation σ.

Looked at from this point of view, the quantity nVB which we have been referring to as the variation between the samples can be considered as a measure of variation between the treatments. Traditionally, then, the quantity nVB has been called the "treatment sum of squares" and has often been denoted by the symbols SST. What about the quantity VW? This number, which we have been calling the variation within the samples, represents the error made in using the mean of each sample as an estimate of the true mean of the corresponding population. Historically, then, VW has been called the "error sum of squares" and has often been denoted by the symbol SSE. In the language of treatment sum of squares and error sum of squares, the F statistic used to test the hypothesis in question would be

$$F = \frac{SST/(m - 1)}{SSE/(N - m)}$$

where m = number of samples of data

N = total number of data points

$$SST = \sum\left(\frac{S^2}{n}\right) - \frac{(\Sigma x)^2}{N}$$

$$SSE = \Sigma x^2 - \sum\left(\frac{S^2}{n}\right)$$

We then reject **H**, concluding that the treatment effects differ among themselves, at level α if $F > F_\alpha[m - 1, N - m]$, i.e., if the treatment sum of squares is substantially larger than the error sum of squares.

EXERCISES 7.B

1   In the situation of Exercise 7.A.1, use the procedures of one-way analysis of variance to decide at level $\alpha = .05$ whether or not there are significant differences between the monthly premiums of the four competing plans.

2   Carry out a one-way analysis of variance based on the data of Exercise 7.A.2 in order to decide at level $\alpha = .01$ whether or not there are significant differences in aptitude among persons in the six occupational classifications.

3   Apply the one-way analysis of variance technique to the ichthyological data of Exercise 7.A.3 to find out at level $\alpha = .05$ whether temperature changes affect the weights of fish.

4   Using the data of Exercise 7.A.4, run a one-way analysis of variance to answer the question whether the average appropriations are basically the same to each of the three subdivisions. Use a level $\alpha = .01$.

5   A New Jersey tomato farmer wants to find out if his land is more suitable for production of one kind of tomato rather than another, or if all kinds of tomatoes grow equally well on his land. He conducts an experiment planting beefsteak tomatoes on nine plots, cherry tomatoes on seven plots, and pear tomatoes on six plots. The resulting yields, in pounds of tomatoes, follow:

Tomato Yield, pounds		
Beefsteak	Cherry	Pear
120	160	110
110	140	90
140	130	100
100	140	120
120	150	100
110	140	80
130	140	
140		
130		

At level $\alpha = .05$, do the results of the experiment indicate the existence of significant differences among the yields of the three kinds of tomatoes?

6   A psychologist's research assistant wants to find out if porpoises, monkeys, and rats learn at the same rate of speed. She places one animal at a time in an appropriate T maze, the left exit of which contains a "punishment," such as an electric shock, and the right exit contains a "reward," such as food. The assistant runs 6 porpoises, 9 monkeys, and 11 rats through the maze,

and she records the number of times through that it takes each animal to learn that the right fork is the one to take. The learning times for each animal are as follows:

Porpoises	3	7	9	15	6	2					
Monkeys	18	14	7	5	11	9	3	11	12		
Rats	4	21	11	16	19	23	7	10	17	6	8

At level $\alpha = .05$, can she conclude that all three species of animals have the same average learning times?

7 Accident figures accumulated by the National Safety Council over various 3-day weekend holidays show the following number of fatalities in each of three cities of comparable size:

Number of Fatalities		
Baltimore	Detroit	San Diego
20	22	16
18	21	14
25	28	15
28	30	20
29	29	25

At level $\alpha = .05$, do the data indicate that the number of fatalities differs significantly among the three cities?

SUMMARY AND DISCUSSION

In the present chapter, we have extended Chap. 4's discussion of two-sample tests for the difference of means to the case of more than two samples. Using one-way analysis of variance, we were able to consider several samples simultaneously, testing the hypothesis that all their population means are equal against the alternative that significant differences exist among the means. The statistical analysis was based on the relationships between the variation among the sample means and the variation of the data points within each sample.

SUPPLEMENTARY EXERCISES

1 In a study of whether or not various kinds of pollutants in the air inhibit the growth of mice, 28 mice born at the same time were fed exactly the same diet in order to prepare them for the experiment. All other living conditions were

identical, except that the mice were divided into four groups according to the type of pollution in their atmosphere. A biologist participating in the study collected the following data on weight gains of the mice over a period of time:

Weight Gains of Mice, grams			
Atmosphere with Cigarette Smoke	Atmosphere with Auto Exhaust	Atmosphere with Industrial Smoke	Clean Air
6 *36*	9 *81*	10 *100*	11 *121*
7 *49*	7 *49*	9 *81*	8 *64*
6 *36*	9 *81*	8 *64*	10 *100*
10 *100*	8 *64*	5 *25*	9 *81*
8 *64*	5 *25*	10 *100*	8 *64*
5 *25*	10 *100*	8 *64*	6 *36*
7 *49*	8 *64*	6 *36*	11 *121*

49 *56* *56* *63 = 224*
9 = 8

(a) Calculate the components of the variance.
(b) Using the components of the variance, decide at level $\alpha = .05$ whether or not there are significant differences in weight gains of mice in the four types of atmospheres.
(c) Carry out a one-way analysis of variance to answer the question of part b.

2 It is the job of a tea tester for an English tea-importing company to decide which tea leaves are most suitable for English Breakfast tea. The tea tester has to decide between three lots of leaves, an inexpensive lot, a moderately priced lot, and an expensive lot. The tea tester and his laboratory staff discover the following levels of impurities in eight randomly selected leaves of each lot:

Inexpensive lot	3	4	25	9	11	4	12	20	*88* *11*
Moderately priced lot	4	9	16	9	12	6	4	8	*68* *8.5*
Expensive lot	6	5	7	10	6	8	5	4	*51* *6.4*

8.6

(a) Calculate the components of the variance.
(b) Use the components of the variance to test at level $\alpha = .01$ whether or not all three lots of leaves have the same average impurity levels.
(c) Apply the one-way analysis of variance technique to test the same hypothesis as in part b.

3 Peanuts are known to be an excellent source of protein, and they are therefore a good substitute for meat and fish when prices are high. To find out whether storage of peanuts for a length of time tends to change their protein content, a nutritionist selected several bags of peanuts from lots that had been in storage for various periods of time, and then she measured the protein content of each bag. The data, in grams of protein per bag, follow:

Fresh	Stored 6 Months	Stored 12 Months	Stored 18 Months	Stored 24 Months
10 100	8 64	5 25	10 100	7 49
8 64	6 36	8 64	6 36	8 64
6 36	5 25	9 81	9 81	7 49
10 100	10 100	8 64	10 100	11 121
8 64	7 49	8 64	6 36	5 25
10 100	9 81	6 36	8 64	9 81
6 36	6 36	10 100	7 49	6 36
8 64		5 25	7 49	8 64
6 36		7 49		10 100
10 100		5 25		
4 16				

86 51 71 63 71 = 342

At level $\alpha = .05$, do the data indicate that storage of peanuts tends to alter their protein content?

4   A factory, in the process of assessing its energy costs, conducted a study of how much it costs to run machines with each of five different energy sources. After randomly assigning machines to energy sources, careful monitoring yields the following monthly costs, in hundreds of dollars:

Energy Source				
Electricity	Natural Gas	Oil	Coal	Manual
6 36	5 25	8 64	4 16	5 25
4 16	3 9	6 36	2 4	4 16
3 9	3 9	4 16	2 4	2 4
2 4	3 9	3 9	3 9	6 36

15   14   21   11   17 = 78

At level $\alpha = .01$, do the data provide evidence sufficient to assert that different energy sources result in different operating costs on the average for the machines under study?

5   An economist for HEW is conducting a study of vegetable prices in various metropolitan areas for the purpose of estimating the cost to an individual of maintaining a diet containing all the recommended daily amounts of nutrients. The following data give the retail prices per pound of four basic vegetables in randomly selected areas:

Lettuce	Tomatoes	Carrots	Onions
30 900	28 784	34 1156	26 676
18 324	25 625	17 289	17 289
28 784	21 441	19 361	31 961
26 676	30 900	32 1024	29 841
20 400	29 841	30 900	19 361
25 625	16 256	27 729	20 400
26 676	31 961	28 784	31 961

173   180   187   173 = 713

At level $\alpha = .05$, does the study indicate that there are substantial price differences between the various vegetables listed?

6 As part of a study of the role of the coffee break in the proper functioning of the American economy, a corporation psychologist selected 20 employees of a major company, classified them according to number of cups of coffee drunk per day, and rated each on a job performance scale. The ratings are:

No Cups	1-3 Cups	4-6 Cups	7-9 Cups	10 or More Cups
15	8	12	18	12
10	9	13	12	12
9	11	8	17	15
11	14	15	17	18

At level $\alpha = .01$, do differences in the number of cups of coffee consumed lead to significant differences in job performance ratings?

NONPARAMETRIC TESTS

In most of the applied situations studied in the previous chapters using the format of testing statistical hypotheses, there were somewhat stringent requirements on the nature of the data involved. The small-sample ($n < 30$) tests, for example, could not be applied unless it was reasonable to assume that the underlying populations from which the random samples were drawn were normally distributed. The restriction to normally distributed data was needed because an integral part of the testing procedure made use of a table derived from the normal distribution, namely, that of the t distribution. Statistical tests falling into this category were the t tests of means in Chap. 4 and the correlation test for linearity in Chap. 5. Our objective in this chapter is to explain how to handle applied situations in which the prerequisites for using these tests are not met.

All the statistical tests studied in Chaps. 4, 5, and 7 dealt with *parameters* of populations. The concept of a parameter was discussed in some detail in Chap. 2 in connection with the binomial and normal distributions. It was pointed out there that both of these distributions possess certain characteristic numbers, called "parameters," from which the complete behavior pattern of the distribution can be determined. For the binomial distribution, the parameters were denoted by n and p, while for the normal distribution they were called μ and σ. The tests of Chaps. 4 and 7 all involved the study of one or another of these parameters: We had one- and two-sample (and more-sample) tests for population means μ (of normal distributions) and one- and two-sample tests for population proportions p (of binomial distributions). Because these tests all involve parameters, they can be referred to as "parametric" tests.

If the underlying populations from which our random samples were selected do not have distributions which behave in accordance with the table of either of the normal or binomial distribution, then these populations cannot be characterized by parameters μ or p. It logically follows that applied problems involving averages of non-normal populations cannot be solved[1] using the usual parametric tests of μ because such populations do not possess parameters having the same statistical properties as μ. Nevertheless, small-sample problems involving non-normal populations often arise in applied contexts, and so it is necessary to develop statistical techniques for analyzing and solving these problems. For reasons that should now be obvious, these new techniques are called "nonparametric."

The correlation test (appearing in Chap. 5) of linearity between two sets of data, the x's and the y's, is also a parametric test. The parameter involved is the "true correlation" ρ relating the two normally distributed populations of x's and y's. The correlation ρ is a parameter of the "bivariate normal distribution," a two-dimensional probability distribution formed by the intertwining of two normal populations. In the non-normal case, we use a nonparametric replacement, the "Spearman test of rank correlation." (The word "rank" is not used here in a derogatory sense, but merely refers to the mathematical methods used.)

SECTION 8.A
THE SIGN TEST

When we would like to use the one-sample t test or the paired-sample t test on an applied problem, but the population from which the data points were selected is probably not normally distributed, the sign test provides a valid way of analyzing the situation. While the sign test has the valuable advantage that it can be used in cases involving non-normal data, it suffers from a defect common to all nonparametric tests—the one-sample sign test, for example, pays no attention at all to the actual numerical values of the data points, but merely looks whether they are each below or above the number which plays the role of the hypothesized mean. This makes it a test of whether or not the true median is equal to the hypothesized median, rather than a test of whether the true mean is equal to the hypothesized mean. In fact, the sign test can be viewed as a particu-

---

[1]Unless there are enough data points available to satisfy the criteria of the central limit theorem.

lar type of test for the binomial proportion $p = 1/2$, where the two classifications for the binomial distribution are the categories *above* and *below* the hypothesized median. Small-sample versions of the sign test base their rejection of the hypothesis on the table of binomial distribution, while large-sample versions use the normal approximation to the binomial.

ONE-SAMPLE SIGN TEST

To point up very clearly the distinctions between the parallel parametric and nonparametric tests, we illustrate the sign test by first using it to analyze the same situation we faced in the comparable section of Chap. 4.

Example 8.1 Body Temperatures A medical doctor who is also an amateur anthropologist is interested in finding out whether the average body temperature of Alaskan Eskimos is significantly lower than the usual American average, which is 98.6 degrees. Eight Eskimos selected at random from the state of Alaska census lists had the following recorded body temperatures in degrees:

| 98.5 | 98.1 | 98.6 | 98.7 | 98.4 | 98.9 | 98.0 | 98.4 |

At significance level $\alpha = .05$, do the results of the study support the assertion that Eskimos really have lower body temperatures than Americans who are natives of warmer climates?

SOLUTION   If we denote by μ^* the true median body temperature of Alaskan Eskimos, the sign test is capable of testing the hypothesis

H: $\mu^* = 98.6$ (Eskimos have the same average body temperature as other Americans)

against the alternative

A: $\mu^* < 98.6$ (Eskimos have lower body temperatures)

The number 98.6 here is called the "hypothesized median," by analogy with the hypothesized mean appearing in the one-sample tests of Chap. 4, and we use the symbol

$$\mu_0^* = \text{hypothesized median}$$

The preliminary computational setup for the one-sample sign test appears in Table 8.1. Major components of the actual test for the rejection of **H** in favor of **A**

TABLE 8.1
Calculations for the one-sample sign test
(Body temperature data)

x	μ_0^*	Sign of $x - \mu_0^*$
98.5	98.6	−
98.1	98.6	−
98.6	98.6	0
98.7	98.6	+
98.4	98.6	−
98.9	98.6	+
98.0	98.6	−
98.4	98.6	−

Total Number of +'s = 2

$\theta = 2$
$n = 8 - 1 = 7$
$p = \frac{1}{2}$

are the quantities

θ = number of data points exceeding the hypothesized median μ_0^*

= number of plus signs in the list of signs

n = number of data points not equaling μ_0^*

= number of pluses, plus the number of minuses

$$p = \frac{1}{2}$$

The numbers n and p are used in the small-sample case as the parameters of the appropriate binomial distribution. For a test such as the sign test, which involves the median, we always use $p = 1/2$ because, if the hypothesis **H** is really true, we should ideally have half pluses and half minuses in the list of signs.

From the calculations of Table 8.1, we find out that we have to use the binomial distribution with parameters $n = 7$ and $p = 1/2$. The portion of Table A.2 which is reproduced in Table 8.2 indicates that, if we use a significance level of $\alpha = .0625$ instead of $\alpha = .05$, we can reject

H: $\mu^* = 98.6$ in favor of **A:** $\mu^* < 98.6$

if $\theta < 2$.

We derive this rejection rule from the fact that

$$P(\theta < 2) = P(\theta \leq 1) = .0625$$

TABLE 8.2
A small portion of Table A.2

n	k	$p = .50$
7	0	.0078
	1	.0625
	2	.2266

according to Table 8.2 if θ is a binomial random variable with parameters $n = 7$ and $p = 1/2$.

As it has turned out, however, we have obtained from the calculations in Table 8.1 that $\theta = 2$. And if $\theta = 2$, then it is not true that $\theta < 2$. Therefore we cannot reject **H** at level $\alpha = .0625$. Because .05 is between .0078 and .0625, this decision also implies that we cannot reject **H** at level $\alpha = .05$. Our conclusion, then, at level $\alpha = .05$ is that the data do not support the assertion that Eskimos have lower body temperatures on the average than Americans of warmer climates.

Our decision, in regard to the question of Example 8.1, to support the assertion that Eskimos do not have lower body temperatures than Americans of warmer climates was the same decision to which we came in Example 4.2 using the one-sample t test. Although it is reassuring that this happened, it is frequently the case that parametric and nonparametric tests applied to the same set of data will recommend "opposite" decisions. In the case of the sign test versus the t test, this disparity could occur because the true median, for example, is larger than the hypothesized number, while the true mean is not. The decisions are not really opposite because both tests do not really test exactly the same thing, but the appearance of opposite is given if we think about the problem in terms of average rather than the more precise terms mean and median. In case of an extreme difference between the mean and the median (a possibility we pointed out in Chap. 1), it should not be surprising if the t test and the sign test recommend different decisions. Which test is to be believed in the case of opposite decisions? Well, if the data come from a normally distributed population, the parametric test takes precedence because it makes use of the data directly rather than of "shadow" properties of the data, such as signs. On the other hand, if the normality or other conditions for the parametric test are not met, then the nonparametric test dominates because the parametric test simply does not apply.

One more comment on the one-sample sign test: If $n > 20$, then the binomial table of the Appendix gives way to the normal approximation to the binomial distribution, by analogy with the large-sample test of proportion discussed in Sec. 4.E. We then base our decision to reject or accept **H** on a z test, where

$$\bigstar \quad z = \frac{\theta - (n/2)}{\sqrt{n/4}} \tag{8.1}$$

This formula is a consequence of formula (4.7) of the large-sample test of proportion where $\hat{p} = \theta/n$ and $p_0 = 1/2$. This large-sample formula can also be used for the paired-sample sign test.

PAIRED-SAMPLE SIGN TEST

Example 8.2 Cloud Seeding A meteorologist participating in a project to determine to what extent, if any, mankind can influence local weather conditions has set up an experiment to test the effectiveness of present methods of cloud seeding in the artificial production of rainfall. Two farming areas in Nebraska with similar past meteorological records, lying 150 miles apart in a north-south direction, were selected for the experiment. The Fairbury area is regularly seeded throughout the year, while the Norfolk area is left unseeded. Their monthly precipitation is recorded in Table 8.3. At level $\alpha = .01$, do the results of the experiment indicate that cloud seeding significantly increases monthly precipitation?

SOLUTION We apply the paired-sample sign test as a nonparametric substitute for the paired-sample t test. If we use the notation

$\mu_d^* =$ true median of monthly differences in precipitation (unseeded region minus seeded region)

then we want to test the hypothesis

H: $\mu_d^* = 0$ (cloud seeding does not increase precipitation)

TABLE 8.3
Nebraska precipitation data from cloud-seeding experiment

Month	Inches of Precipitation	
	Norfolk (Unseeded)	Fairbury (Seeded)
Jan.	1.5	1.4
Feb.	1.4	1.4
Mar.	2.2	2.6
Apr.	2.6	2.5
May	3.9	4.8
June	4.5	4.3
July	3.7	4.0
Aug.	3.4	3.5
Sept.	4.0	3.9
Oct.	2.5	2.6
Nov.	1.9	1.7
Dec.	1.5	1.4

TABLE 8.4
Calculations for the paired-sample sign test
(Cloud-seeding data)

Month	Inches of Precipitation Unseeded Area x_U	Seeded Area x_S	Sign of $d = x_U - x_S$
Jan.	1.5	1.4	+
Feb.	1.4	1.4	0
Mar.	2.2	2.6	−
Apr.	2.6	2.5	+
May	3.9	4.8	−
June	4.5	4.3	+
July	3.7	4.0	−
Aug.	3.4	3.5	−
Sept.	4.0	3.9	+
Oct.	2.5	2.6	−
Nov.	1.9	1.7	+
Dec.	1.5	1.4	+

Total Number of +'s = 6

$\theta = 6$
$n = 12 - 1 = 11$
$p = \frac{1}{2}$

against the alternative

A: $\mu_d^* < 0$ (cloud seeding increases precipitation)

Table 8.4 contains the calculations necessary to carry out the statistical analysis. As Table 8.4 indicates, the alternative is formulated as $\mu_d^* < 0$ because a negative value of μ_d^* represents more precipitation in the seeded region.

The results of the calculations done in Table 8.4 show that we should base our statistical test on the binomial distribution having parameters $n = 11$ and $p = 1/2$. From the portion of Table A.2 that is reproduced in Table 8.5, we then see that we should reject **H** in favor of **A** at level $\alpha = .0059$ if $\theta \leq 1$ or $\theta < 2$ and at level $\alpha = .0327$ if $\theta \leq 2$ or $\theta < 3$. Because $\theta = 6$, and $.0059 < .01 < .0327$, we cannot reject **H** at level $\alpha = .01$. Our conclusion at level $\alpha = .01$ is that cloud seeding does not seem to significantly increase rainfall, according to the given data.

TABLE 8.5
A small portion of Table A.2

n	k	$p = .50$
11	1	.0059
	2	.0327

EXERCISES 8.A

1 A psychologist, conducting a study of the average person's ability to judge distances, sets up a test of depth perception in which randomly selected individuals attempt to estimate the distance between two markers. The markers were actually 2.5 feet apart, while the 10 participants in the study gave the following estimates:

2.1	1.8	2.3	2.3	2.6	2.5	2.3	2.5	2.1	2.5

Use the sign test at level $\alpha = .05$ to find out whether or not the results of the study indicate that the average person has difficulty in estimating the correct distance of 2.5 feet.

2 In Supplementary Exercise 3 of Chap. 4, a company which markets canned carrots advertises that their 15-ounce cans actually weigh 15.1 ounces on the average. From the random sample of 11 weights listed in that exercise, can we conclude at level $\alpha = .05$ that
   (a) the true median contents are below the 15.1 ounces claimed by the company?
   (b) the true median contents are below the 15.0 ounces printed on the label?

3 It is the policy of one school district to hire credentialed reading specialists whenever the true median reading score of the district's sixth graders falls below 40, as measured by a particular standard test. Of a random sample of 25 pupils, 18 had scores below 40 and the other 7 had scores above 40. Is the hiring of reading specialists justified
   (a) At significance level $\alpha = .05$?
   (b) At significance level $\alpha = .005$?

4 From the data of Exercise 4.D.1, does the sign test support at level $\alpha = .05$ the theory that employees have less absenteeism after they stop smoking?

5 From the data of Exercise 4.D.6, does the sign test support at level $\alpha = .05$ the assertion that the kindergarten experience seems to help first graders?

6 Ten sets of identical twins were administered drugs intended to increase the pulse rate in an experiment designed to determine if identical twins have similar reactions to medication. In four of the ten sets, the older twin had a larger increase in pulse rate, while in the remaining six sets, the younger twin had the larger increase. At level $\alpha = .10$, does the experiment tend to support the hypothesis that there is no general rule concerning which twin, the older or the younger, has the larger increase?

SECTION 8.B
THE SPEARMAN TEST OF RANK CORRELATION

As we have noted during the course of our discussion of the sign test, we lose some control over the problem when we convert the original data points into shadow properties such as signs. The unpleasant consequences of the conversion can be best seen in the fact that we cannot use the sign test to check on the differences of means but rather on the differences of medians. A similar situation occurs in connection with the "Spearman test of rank correlation," which requires us to rank each set of data in numerical order and then to compute the correlation coefficient of the ranks. It turns out that we cannot use the Spearman test to check on the existence of a linear relationship between two sets of data but only on the existence of a monotone relationship. By a "monotone" relationship between x and y, we mean a relationship in which y increases as x increases, or y decreases as x increases. Examples of monotone *increasing* relationships can be seen graphically in Figs. 5.3, 5.6, 5.9*b*, and 5.11. Monotone *decreasing* relationships are illustrated in Figs. 5.1 and 5.8. All linear relationships are monotone, but, as Fig. 5.9*b* shows, not all monotone relationships are linear. On the other hand, the parabolic data of Fig. 5.9*a* express a relationship which is not monotone because as x increases, y starts to decrease but, after a while, turns upward and begins to increase.

Because the Spearman test uses only the ranks of the data points and not their original numerical values, it is not sharp enough to be able to judge the degree of linearity. On the other hand, the Spearman test has two distinct advantages: (1) Its usage is valid in the presence of non-normal data. (2) It works excellently in situations where at least one of the two data sets consists entirely of rankings rather than measurements. One widely known example of data which consists entirely of rankings is the weekly (during the football season) ranking of the top 10 college teams. A rank correlation problem here would be to test the correlation between the national rankings of the top 10 teams (ranked data) and the total weight of their seven first-string linemen (measurement data). It makes no sense to apply the Pearson test of linearity to this question, but the Spearman test fits quite well.

LARGE SAMPLES

Example 8.3 Control of Fever by Aspirin Substitute Because a significant fraction of the general population is known to be allergic to aspirin, the need for an effective substitute is evident. In the course of the laboratory development of a possible substitute DAT-1, one seriously ill patient is given a dose every 4 hours for 48 hours, and his body temperature is recorded 2 hours after each dosage. His temperature, as time passes, is recorded in Table 8.6. At level $\alpha = .01$, can we assert with confidence that a person's body temperature decreases as we continue to administer the drug DAT-1 every 4 hours?

SOLUTION We are asking whether or not temperature decreases as hour of dosage increases. It is therefore appropriate to test whether or not temperature and hour of dosage are related in a monotone *decreasing* manner, and so the

TABLE 8.6
Periodic body temperatures of patient undergoing aspirin substitute treatment

Hour of Dosage	Temperature 2 Hours After Dosage
0	105.7
4	104.2
8	103.4
12	100.6
16	102.4
20	101.2
24	100.1
28	99.7
32	98.9
36	98.8
40	99.5
44	98.7
48	98.6

coefficient of rank correlation will be useful to us. We present the scattergram of the aspirin substitute–temperature data in Fig. 8.1.

We want to test the hypothesis

H: Relationship between temperature and hour of dosage is not monotone

FIGURE 8.1
Scattergram of aspirin substitute–temperature data.

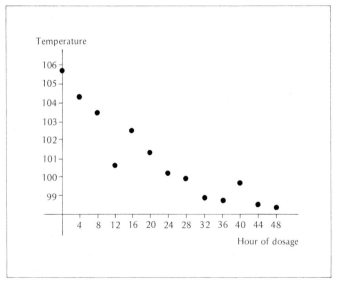

TABLE 8.7

Rejection rules in the presence of various alternatives for the Spearman test of rank correlation

(**H**: the relationship is not monotone)

Alternative	*Reject **H** in Favor of **A** at Significance Level α if:*
A : Monotone increasing	$z > z_\alpha$
A : Monotone decreasing	$z < -z_\alpha$
A : Monotone	$|z| > z_{\alpha/2}$

NOTE: The use of z is valid only in the large-sample case when there are more than 10 pairs of data points. For the small-sample case, special tables must be used.

against the alternative

> **A**:   Relationship is a monotone decreasing one

To test **H** against **A**, we first determine

> n = number of pairs of data points

If $n > 10$, then a modified version of the central limit theorem allows us to use a z test, where

★ $$z = r_S \sqrt{n - 1} \qquad (8.2)$$

Here r_S = Spearman coefficient of rank correlation. The rejection rule is as follows, selected appropriately from Table 8.7: We reject **H** in favor of **A** at level $\alpha = .01$ if

$$z < -z_{.01} = -2.33$$

All that now remains is the computation of z, and the bulk of this computation is concerned with the calculation of r_S. We carry out the calculation of r_S in Table 8.8 using the formula

★ $$r_S = 1 - \frac{6\Sigma d^2}{n(n^2 - 1)} \qquad (8.3)$$

The operative term in the formula for r_S is the quantity Σd^2, the sum of the squared differences of the *ranks* of the data points in each pair. As shown in Table 8.8, we rank the data points in each column separately, lowest to highest, we compute the pairwise differences d between the *ranks* of x and y, and we

TABLE 8.8
Calculations leading to the Spearman coefficient of rank correlation
(Aspirin substitute–temperature data)

Hour of Dosage, x	Rank r_x	Temperature 2 Hours after Dosage, y	Rank r_y	$d = r_x - r_y$	d^2
0	1	105.7	13	−12	144
4	2	104.2	12	−10	100
8	3	103.4	11	−8	64
12	4	100.6	8	−4	16
16	5	102.4	10	−5	25
20	6	101.2	9	−3	9
24	7	100.1	7	0	0
28	8	99.7	6	2	4
32	9	98.9	4	5	25
36	10	98.8	3	7	49
40	11	99.5	5	6	36
44	12	98.7	2	10	100
48	13	98.6	1	12	144
Sums				0	716

$n = 13$

$$r_s = 1 - \frac{6\Sigma d^2}{n(n^2 - 1)} = 1 - \frac{6(716)}{13(13^2 - 1)} = 1 - \frac{4296}{(13)(169 - 1)} = 1 - \frac{4296}{(13)(168)}$$
$$= 1 - 1.967 = -.967$$

then come up with Σd^2. From the calculations in Table 8.8, we get

$$r_s = 1 - \frac{6\Sigma d^2}{n(n^2 - 1)} = 1 - \frac{6(716)}{13(13^2 - 1)}$$

$$= 1 - \frac{4296}{(13)(168)} = 1 - 1.967 = -.967$$

It follows at once that

$$z = r_s \sqrt{n - 1} = -.967\sqrt{13 - 1} = -.967\sqrt{12} = (-.967)(3.464)$$
$$= -.3.35$$

We have agreed to reject **H** in favor of **A** if $z < -2.33$. Since it turned out that $z = -3.35 < -2.33$, we do, in fact, reject **H**. At level $\alpha = .01$, we therefore conclude that temperature is monotone decreasing as the hours of dosage with the aspirin substitute DAT-1 pass. The data does contain evidence for the assertion that the periodic use of DAT-1 tends to reduce body temperature.

SMALL SAMPLES

When fewer than 11 pairs of data points are available, it is not valid to use the percentage points of the normal distribution for the Spearman test of rank correlation. We must refer instead to a special table (Table A.7 of the Appendix) giving the levels of significance for small-sample tests of rank correlation. The following example, based on the parabolic data of Fig. 5.9a and Table 5.4a, illustrates the small-sample procedure.

Example 8.4 Psychological Learning Theory A psychologist wants to find out if she can teach color differentiation to monkeys by using bananas as rewards. She sets up a row of several balls of different colors, hands the monkeys a blue ball and several bananas, and asks the monkeys to choose the ball of the same color from the row. She records the number of bananas given to and the number of mistakes made by each of the monkeys. The resulting data is presented in Table 8.9. At level $\alpha = .05$, do the results of the experiment indicate that the number of mistakes made by a monkey bears a monotone relationship with the number of bananas it received?

SOLUTION We apply the Spearman test of rank correlation to test the hypothesis

> **H**: The relationship is not monotone

against the alternative

> **A**: The relationship is monotone

According to the information on rejection rules presented in Table 8.7 (adjusted mentally to the small-sample case), we compute r_s and we reject **H** at level α if

$$|r_s| > r_s[n, \alpha/2]$$

TABLE 8.9
Psychological learning data

Monkey	Number of Bananas	Number of Mistakes
A	8	9
B	1	16
C	6	1
D	3	4
E	5	0
F	2	10

TABLE 8.10
A small portion of Table A.7

N		Values of Σd^2 6
6	α	.029
	r_s	.83

Here $n = 6$, and Table A.7 contains the information that

$$r_s[6, .029] = .83$$

(The relevant portion of Table A.7 appears in Table 8.10.) As we have found, when working with the small-sample sign tests, we cannot always conduct a test at the exact significance level desired. However, we can usually do it at a very close level. In this case, we want to conduct the test at level $\alpha = .05$; however, the nature of r_s allows us to have $\alpha/2 = .029$ rather than .025, and so we must use a level $\alpha = (.029)(2) = .058$. At level $\alpha = .058$, then, we would reject **H** in favor of **A** if

$$|r_s| > r_s[6, .029] = .83$$

It remains now only to compute r_s for the data of Table 8.9. We do the computations in Table 8.11, and they show that

$$r_s = 1 - \frac{6\Sigma d^2}{n(n^2 - 1)} = 1 - \frac{6(56)}{6(6^2 - 1)} = 1 - \frac{336}{(6)(36 - 1)} = 1 - \frac{336}{(6)(35)}$$

$$= 1 - 1.6 = -.6$$

Therefore $|r_s| = .6$, which does not exceed our rejection level of $r_s[6, .029] = .83$. Consequently, we cannot reject **H** at level $\alpha = .058$. Our conclusion is that the relationship between number of mistakes made by the monkeys and number of bananas given to them is not monotone. There seems to be no significant trend in either direction, monotone increasing or monotone decreasing.

One final remark: If there are two or more data points which have the same numerical value and therefore have tied ranks, we divide up their rank positions equally. For example, if the third, fourth, and fifth ranking data points were all equal, they would each be assigned a rank of $(3 + 4 + 5)/3 = 12/3 = 4$. Unless there are an unusually large number of ties, no adjustment of the formula for r_s is necessary.

ESSENTIALS
OF STATISTICS

TABLE 8.11

Calculations leading to the Spearman coefficient of rank correlation
(Psychological learning data)

Monkey	Number of Bananas, x	Rank r_x	Number of Mistakes, y	Rank r_y	$d = r_x - r_y$	d^2
A	8	6	9	4	2	4
B	1	1	16	6	−5	25
C	6	5	1	2	3	9
D	3	3	4	3	0	0
E	5	4	0	1	3	9
F	2	2	10	5	−3	9
Sums					0	56

$n = 6$

$$r_s = 1 - \frac{6\Sigma d^2}{n(n^2 - 1)} = 1 - \frac{6(56)}{(6)(6^2 - 1)} = 1 - \frac{336}{(6)(36 - 1)} = 1 - \frac{336}{(6)(35)} = 1 - 1.6 = -.6$$

EXERCISES 8.B

1 Referring to the data of Exercise 5.C.7,
   (a) at level $\alpha = .05$, does one's job performance rating seem to be related in a monotone way with one's math aptitude score?
   (b) Draw a scattergram of the data.
2 Based on the data of Exercise 5.C.5,
   (a) does the total cost of production seem to increase monotonically at level $\alpha = .01$ as the quantity produced increases?
   (b) Draw a scattergram of the data.
3 A geographer would like to find out whether or not the amount of cropland in a region bears a monotone relationship to the amount of lowland in that region. The following data compare the percentage of lowland in each of 20 randomly selected counties across the United States with the percentage of cropland in the county:

County	Percentage of Lowland	Percentage of Cropland
A	48	28
B	62	45
C	35	22
D	49	35
E	100	75
F	88	62
G	0	3
H	41	24
I	40	21

(Continued on next page)

County	Percentage of Lowland	Percentage of Cropland
J	40	21
K	61	25
L	68	26
M	33	17
N	0	4
O	100	70
P	37	29
Q	44	36
R	90	73
S	100	82
T	100	84

(a) At level $\alpha = .05$, do the data indicate that the percentage of cropland is related in a monotonically increasing manner with the percentage of lowland?

(b) Draw a scattergram of the data.

4 Twelve persons selected at random are asked the following two questions by a psychologist as part of a job satisfaction study: (1) "What is your hourly pay?" (2) "How would you rate your job satisfaction on a scale of 0 (low) to 10 (high)?" The results of the survey are presented below:

Respondent	A	B	C	D	E	F	G	H	I	J	K	L
Hourly pay, dollars	8	2	6	4	4	20	10	6	6	4	11	15
Job satisfaction	6	4	5	4	3	8	7	4	1	2	9	5

(a) At level $\alpha = .05$, does there seem to be a monotone relationship between job satisfaction and hourly pay?

(b) Draw a scattergram of the data.

5 As part of a study into the question of whether or not you should judge a book by its cover, an advertising executive hired two groups of consultants to judge each of 10 books according to artistry of the cover and internal literary content. Each book was given a ranking from 1 to 10 for its cover and again for its content. The results follow:

Book	#1	#2	#3	#4	#5	#6	#7	#8	#9	#10
Cover rank	6	4	10	9	7	8	3	2	5	1
Content rank	5	7	1	2	4	3	8	9	6	10

Do the results of the study indicate at level $\alpha = .10$ that you should judge a book by its cover?

6 An analysis of the relation between the net incomes of 16 major banking companies and the total return on investment to their stockholders yielded the following information:

Banking Company	Rank According to Net Income	Rank According to Return to Stockholders
Bank of America	2	9
Citibank	1	12
Chase Manhattan	3	15
J. P. Morgan	4	2
Manufacturers Hanover	5	6
Chemical New York	8	3
Bankers Trust	10	7
Continental Illinois	7	14
First Chicago	6	16
United California	9	11
Security Pacific	12	10
Wells Fargo	13	13
Marine Midland	14	8
Irving Trust	15	4
Crocker National	16	5
Mellon National	11	1

On the basis of the above data (which described the situation for one year in the recent past), could a financial analyst have concluded at level $\alpha = .10$ that the relationship between net income and return to stockholders was monotone?

SUMMARY AND DISCUSSION

In this concluding chapter, we have tied together a number of loose ends which were left hanging in earlier chapters. The small-sample tests of statistical hypotheses that we studied in Chaps. 4 and 5 were based on tables of the t distribution, and this meant that they could be validly applied only to sets of data satisfying stringent conditions. In the present chapter, we have freed ourselves of some of these restrictions by introducing statistical tests based on shadow properties such as signs and ranks of a set of data rather than on the numerical values of the data points themselves. While, on the one hand, we can apply these tests to almost all data regardless of distribution, on the other hand, we get less precise results because we are forced to work with these shadow properties rather than with the actual data points. Working with a nonparametric test is somewhat analogous to studying the shadow of an object for the purpose of drawing conclusions about the object itself. We see the general picture, but we lose sight of the sharp distinctions. When the data do nòt satisfy the required conditions of the parametric tests, however, we have no other choice but to use nonparametric tests because the parametric tests are simply invalid in such contexts. The moral of the story is this: When it is valid to apply a parametric test, use it; when it is not, make the best of whatever nonparametric test is available.

SUPPLEMENTARY EXERCISES

1 A person who feels the need to lose some weight is considering going on one of the two famous diets, the Drink-More diet or the Eat-Less diet, but only if statistical analysis shows that one or both of these diets are effective in inducing weight loss. As it turns out, five acquaintances have tried the Drink-More diet and another five have tried the Eat-Less diet. The weights of the 10 acquaintances, both before beginning and after completing their diet programs, are recorded below:

Drink-More Diet			Eat-Less Diet		
Acquaintance	Weight Before	Weight After	Acquaintance	Weight Before	Weight After
A	150	150	F	150	141
B	160	160	G	160	150
C	170	170	H	170	160
D	180	179	I	180	170
E	190	141	J	190	179

(a) Apply the paired-sample sign test to determine whether or not the hypothesis that the Drink-More diet is not effective can be rejected at level $\alpha = .05$. (At what exact level of significance does the paired-sample sign test reject the hypothesis that the Drink-More diet is not effective?)

(b) Apply the paired-sample sign test to determine whether or not the hypothesis that the Eat-Less diet is not effective can be rejected at level $\alpha = .05$.

(c) At what exact level of significance does the paired-sample sign test reject the hypothesis that the Eat-Less diet is not effective?

2 As part of a study of unemployment and unemployment compensation, an economist noted the following data on the situation in seven industrialized countries:

Country	Percentage of Work Force Unemployed	Percentage of Work Force Covered by Unemployment Compensation
United States	8.4	95
Canada	7.2	99
Great Britain	4.7	80
West Germany	4.5	93
France	3.8	61
Italy	3.4	51
Japan	1.7	45

Does the data support the conclusion that a country which ranks high in percentage of work force covered by unemployment compensation also ranks high in percentage of work force unemployed? Use a level of significance of $\alpha = .01$.

3 A social worker is assigned the job of finding out whether or not a particular

6-month vocational training program is effective in increasing the income of its participants. Fourteen participants are selected at random from a group that is just about to begin training, and their monthly incomes are recorded. Eighteen months later (1 year after completion of the training program) their monthly incomes are again noted. The data (in hundreds of dollars) follow:

Participant	Income at Start of Program	Income 18 Months Later
#1	2.2	3.8
#2	.0	4.1
#3	3.9	4.8
#4	6.0	6.0
#5	1.3	4.1
#6	3.9	5.8
#7	2.3	.0
#8	.0	.0
#9	1.5	.0
#10	4.2	4.7
#11	1.6	4.0
#12	2.4	4.2
#13	3.6	4.1
#14	3.5	.0

At level $\alpha = .05$, decide whether the program yields a significant increase in the monthly income of its participants.

4 A behavioral biologist feels that performance of a laboratory rat on an intelligence test depends, to a large extent, on the amount of protein in the rat's daily diet. To check out the theory, he accumulated the following data after working with 10 rats:

Rat	A	B	C	D	E	F	G	H	I	J
No. of units of protein daily	12	12	10	10	10	3	2	20	20	30
Standard test score	20	40	30	80	50	50	90	30	40	40

(a) Draw the scattergram of the data.
(b) Does the information indicate at level $\alpha = .05$ that the relationship between protein consumed and score attained is monotone?

5 In a study of the efficiency of automobile engines with respect to consumption of gasoline, one subcompact car was operated at various speeds and the fuel consumption (in miles per gallon) was carefully measured. The data are as follows:

Speed, miles/hour	10	20	30	40	50	60
Efficiency, miles/gallon	15	20	30	35	25	20

(a) Draw the scattergram of the data.
(b) At level $\alpha = .10$, do the data seem to imply that the relationship between speed and efficiency is monotone?

ANSWERS TO SELECTED ODD-NUMBERED EXERCISES

Exercises 1.A

1	(a) 1–4	10	(b)	.5–4.5	10	(c) 2.5	10
	5–8	20		4.5–8.5	20	6.5	20
	9–12	10		8.5–12.5	10	10.5	10
	13–16	5		12.5–16.5	5	14.5	5
	17–20	5		16.5–20.5	5	18.5	5

3	1500	3	5	(a) 1–5	39	(b) 3	39
	4500	5		6–10	31	8	31
	7500	18		11–15	10	13	10
	10,500	15		16–20	10	18	10
	13,500	6		21–25	5	23	5
	16,500	0		26–30	3	28	3
	19,500	3		31–35	2	33	2

Exercises 1.B

1	(a) 107	3	(a) 444	7	(a) 7
	(b) 107.5		(b) 460		(b) 4
	(c) 110		(c) No		(c) 4
					(d) 8

Exercises 1.C

5 (a) −6   7 −240   9 39
   (b) −36

Exercises 1.D

1	(a) 11.1	3	(a) 230.1
	(b) 8.83		(b) 230.1
	(c) 11.1		(c) 197.1

Exercises 1.E

1 Yes   3 16%   5 27   7 No   9 1.56%

Exercises 1.F

1 (d) A: 2.03   3 #1 (1.00)
       B: 2.07   #3 (.60)
       C: −2.61   #4 (−.33)
       D: 2.24   #2 (−.50)
       E: −2.32
       F: −1.41
       Applicant D gets the job

Supplementary Exercises—Chapter 1

1	(a)	1–7	22	(b)	4	22	3	(a) 17
		8–14	21		11	21		(b) 17
		15–21	9		18	9		(c) 20
		22–28	4		25	4		(d) 7.15
		29–35	4		32	4		(e) 6

5 (a) 14.35   7 93.75%   9 18,106
   (b) 14
   (c) 6.36

11 51/100 of 1%   13 4%   15 30.9%

Exercises 2.A

1	1–4	.20	3	.0–19.5	.30
	5–8	.40		19.5–39.5	.15
	9–12	.20		39.5–59.5	.05
	13–16	.10		59.5–79.5	.10
	17–20	.10		79.5–100.0	.40

Exercises 2.B

1 (a) $n = 4, p = \frac{1}{2}$

(b) 0, 1, 2, 3, 4

(c)

k	$P(H = k)$
0	.0625
1	.2500
2	.3750
3	.2500
4	.0625

3 (a) $n = 12, p = .2$

(b) 0, 1, 2, . . . , 12

(c) 2.4

(d) 1.39

(e)

k	$P(A = k)$
0	.0687
1	.2062
2	.2834
3	.2363
4	.1328
5	.0532
6	.0155
7	.0033
8	.0005
9	.0001
10	.0000
11	.0000
12	.0000

5 (a) $n = 20, p = .05$

(b) .3585

(c) .3773

(d) .0003

(e) 1.0000

(f) .0003

Exercise 2.C

1 70.16 minutes

3 (a) 3.44%

(b) 91.08%

5 (a) 87.9%

(b) 59.34%

(c) 0.38%

7 A–B: 76.64   D–F: 43.36   9   1136

Exercises 2.D

1 (a) 62.5%

(b) 56.31%

(c) 51.99%

(d) 50.8%

3 (a) 7.55%

(b) .03%

(c) .00%

(d) .00%

5 (a) 16.42%

(b) 1.62%

(c) .00%

(d) .00%

Exercises 2.E

1 (a) .52   3 .0793   5 .0286
    (b) 1.04
    (c) 1.44
    (d) 1.88
    (e) 2.17
    (f) 2.41
    (g) 2.88
    (h) 3.09

Supplementary Exercises—Chapter 2

1 (a) 15.03%   3 .01%   5 (a) 15.13%
    (b) 1.1%   (b) .84%

7 (a) 92.8 minutes   9 86.4   11 (a) 18.7%
    (b) 96.4 minutes   (b) 1.04%
    (c) 103.3 minutes

13   0–57.2   F
    57.3–64.8   D
    64.9–75.2   C
    75.3–82.8   B
    82.9–100.0 A

Exercises 3.B

1 16,182.70–16,266.75   3 (a) 1.97–2.03
                              (b) 1.96–2.04
                              (c) 1.96–2.04

5 14.62–19.38   7 20.95–23.05

Exercises 3.C

1 .70–.83   3 .52–.67   5 (a) .37–.41
                             (b) .37–.41
                             (c) .36–.42

Exercises 3.D

1 57   3 2401   5 4269

Supplementary Exercises—Chapter 3

1 952   3 (a) 2.29–2.71   5 2.19–2.55
          (b) 67

7 (a) 13.42–16.18   9 .12–.18   11 (a) .023–.037
   (b) 12.84–16.76   (b) 27,225

Exercises 4.B

1 $z = -1.8$, yes   3 $z = 2.95$, yes   5 $t = 2.67$, yes

7 $t = .583$, yes

Exercises 4.C

1 $z = 6.53$, yes   3 $t = -2.32$, yes   5 $t = .86$, no

7 $t = 1.07$, no

Exercises 4.D

1 $t = 1.90$, yes   3 $t = 1.68$, yes   5 $t = .62$, no

Exercises 4.E

1 (a) $z = 2$, yes   3 $z = -1.14$, no   5 $z = 1.193$, no
   (b) $z = 6.32$, no

Supplementary Exercises—Chapter 4

1 $z = -4.54$, yes   3 (a) $t = -2.06$, yes   5 $t = 2.059$, yes
                          (b) $t = 0$, no

7 $t = -.79$, no   9 $t = -1.182$, no   11 (a) $z = 1.62$, no
                                                (b) $z = -1.27$, no
13 $z = 1.46$, no

Exercises 5.A

1 (a) $y = 1.75 + .25x$ 3 (a) (0, 6) and (4, 14)
   (b) $y = -2 + 1.5x$ (b) (0, 8) and (3, −1)
   (c) $y = 10 - x$ (c) (0, −3) and (1, 1)

5 $y = 1.47 + .1x$, (0, 1.47)

Exercises 5.B

1 (b) $y = 3.33 + .67x$
(c) (2, 4.67) and (8, 8.67)
(d) 733 dollars

3 (b) $y = -3.6 + 2.8x$
(c) (4, 7.6) and (12, 30)
(d) 18.72

Exercises 5.C

1 92.3%　3 99.5%　5 (b) .723　7 (b) 0%

Exercises 5.D

1 $t = 6.93$, yes　3 $t = 27.93$, yes　5 $t = 3.94$, yes

7 $t = 0$, no

Exercises 5.E

1 (a) 95.6%
(c) $t = -8.07$, yes
(d) $y = 3.23 - .91x$
(e) 1.86

3 (a) 69.8%
(c) $t = -3.04$, yes
(d) $y = 28.8 - 1.22x$
(e) 16.6

5 (a) 57.9%
(c) $t = 2.62$, yes
(d) yes
(e) $y = 22.9 + 6x$
(f) 70,900 dollars
(g) 40,900 dollars

Supplementary Exercises—Chapter 5

1 (a) 88.8%
(c) $t = 6.9$, yes
(d) $y = 1.06 + 5.72x$
(e) 29.66

3 (a) 93.46%
(c) $t = -6.55$, yes
(d) $y = 2148 - 800x$
(e) 388

5 (a) 96.57%
(c) $t = 10.6$, yes
(d) $y = 2.5 + .65x$
(e) 7.05

7 (a) 95.29%
(c) $t = -7.8$, yes
(d) $y = 112.35 - 5.29x$
(e) 33

9 (a) 90.34%
(c) $t = 6.12$, yes
(d) $y = -.65 + 8.7x$
(e) 60.25

Exercises 6.A

1 (c) $W \cup L = \{7, 11, 2, 3, 12\}$
$W \cap L = \phi$
$W^c = \{2, 3, 4, 5, 6, 8, 9, 10, 12\}$
$L^c = \{4, 5, 6, 7, 8, 9, 10, 11\}$
$W^c \cup L^c = S$

$W^c \cap L^c = \{4, 5, 6, 8, 9, 10\}$

$E^c = D$

$E^c \cap D = D$

$E^c \cup D = D$

3 (a) $S = \{3, 4, 5, \ldots, 17, 18\}$ (c) $L \cup E = \{3, 4, 6, 8, 10, 13, 14, 16, 18\}$

                                                     $(L \cup E)^c = \{5, 7, 9, 11, 15, 17\}$

                                                     $L^c = \{5, 6, 7, 8, 9, 10, 11, 14, 15, 16, 17\}$

                                                     $E^c = D$

                                                     $L^c \cap E^c = \{5, 7, 9, 11, 15, 17\}$

Exercises 6.B

1 (a) 1/6   (g) 1/18   3 (a) 1/2

    (b) 1/18   (h) 5/9   (b) 5/8

    (c) 2/9   (i) 0   (c) 1/8

    (d) 1/9   (j) 1/3   (d) 3/8

    (e) 1/2   (k) 0

    (f) 1/2   (l) 1

Exercises 6.C

1 (a) 4/9   3 (a) 3/10   5 6.4%

    (b) 1/9   (b) 2/5   7 44%

    (c) 1   (c) 1/10   9 59%

    (d) 1/2   (d) 7/10

    (e) 0

    (f) 1/9

Exercises 6.D

1 (a) .015625   3 (a) .45   5 28.98%

    (b) .09375   (b) .45

    (c) .234375   (c) .05

    (d) .3125

    (e) .234375

    (f) .09375

    (g) .015625

Exercises 6.E

1 Yes   3 Yes   5 Yes   7 No

Supplementary Exercises—Chapter 6

1 (a) True   3 (a) 66.5%   7 72%
   (b) True   (b) 61.9%
   (c) False
   (d) False

9 55%   11 68%   13 62.5%

15 21.4%   17 Yes   19 Yes

Exercises 7.A

1 (a) $8.90 = 7.02 + 1.88$ 3 (a) $413.25 = 19.25 + 394$
   (b) $F = 24.83$, yes   (b) $F = .39$, no

Exercises 7.B

1 $F = 24.83$, yes   3 $F = .39$, no

5 $F = 18.21$, yes   7 $F = 4.23$, yes

Supplementary Exercises—Chapter 7

1 (a) $88 = 14 + 74$ 3 $F = .36$, no   5 $F = .21$, no
   (b), (c) $F = 1.51$, no

Exercises 8.A

1 They do not   3 (a) $z = -2.2$, yes   5 No
     (b) $z = -2.2$, no

Exercises 8.B

1 (a) $r_s = .027$, no   3 (a) $r_s = .91$, yes   5 $r_s = -1$, yes

Supplementary Exercises—Chapter 8

1 (a) $\alpha = .25$ 3 It does   5 $r_s = .38$, no
   (b) It can
   (c) $\alpha = .0312$

APPENDIX

Chpt. 5   pg 166   Regression

Step 1 - Calculate X, Y, X·Y, X^2, Y^2.
  - sum each colmn
  - find n (number in each colmn)
  - find $r = \dfrac{n\Sigma xY - (\Sigma x)(\Sigma Y)}{\sqrt{n\Sigma x^2 - (\Sigma x)^2}\ \sqrt{n\Sigma Y^2 - (\Sigma Y)^2}}$

  - find r^2 ex/ ~~cost axb~~ $(.942)^2 = .8873 = 88.8\%$

Step 2 - draw scattergram

Step 3 — H: $p=0$ (nonlinear) vs. A $p \neq 0$ (linear)
                                ex/94.2x 2.45
  - Calculate $t = \dfrac{r\sqrt{n-2}}{\sqrt{1-r^2}}$ & reject H
  • if $H > t\alpha_{/2} [n-2]$ (if so it's linear related) Table A.4

Step 4 - $b = \dfrac{n\Sigma xY - (\Sigma x)(\Sigma Y)}{n\Sigma x^2 - (\Sigma x)^2}$

  - $a = \dfrac{\Sigma Y - b\Sigma x}{n}$ & plug into equation
                                         $y = a + bx$

Step 5 - plug # into equation

TABLE A.1
Square roots

N	$\sqrt{N}$	$\sqrt{10N}$	N	$\sqrt{N}$	$\sqrt{10N}$	N	$\sqrt{N}$	$\sqrt{10N}$
1.00	1.000	3.162	1.50	1.225	3.873	2.00	1.414	4.472
1.01	1.005	3.178	1.51	1.229	3.886	2.01	1.418	4.483
1.02	1.010	3.194	1.52	1.233	3.899	2.02	1.421	4.494
1.03	1.015	3.209	1.53	1.237	3.912	2.03	1.425	4.506
1.04	1.020	3.225	1.54	1.241	3.924	2.04	1.428	4.517
1.05	1.025	3.240	1.55	1.245	3.937	2.05	1.432	4.528
1.06	1.030	3.256	1.56	1.249	3.950	2.06	1.435	4.539
1.07	1.034	3.271	1.57	1.253	3.962	2.07	1.439	4.550
1.08	1.039	3.286	1.58	1.257	3.975	2.08	1.442	4.571
1.09	1.044	3.302	1.59	1.261	3.987	2.09	1.446	4.572
1.10	1.049	3.317	1.60	1.265	4.000	2.10	1.449	4.583
1.11	1.054	3.332	1.61	1.269	4.012	2.11	1.453	4.593
1.12	1.058	3.347	1.62	1.273	4.025	2.12	1.456	4.604
1.13	1.063	3.362	1.63	1.277	4.037	2.13	1.459	4.615
1.14	1.068	3.376	1.64	1.281	4.050	2.14	1.463	4.626
1.15	1.072	3.391	1.65	1.285	4.062	2.15	1.466	4.637
1.16	1.077	3.406	1.66	1.288	4.074	2.16	1.470	4.648
1.17	1.082	3.421	1.67	1.292	4.087	2.17	1.473	4.658
1.18	1.086	3.435	1.68	1.296	4.099	2.18	1.476	4.669
1.19	1.091	3.450	1.69	1.300	4.111	2.19	1.480	4.680
1.20	1.095	3.464	1.70	1.304	4.123	2.20	1.483	4.690
1.21	1.100	3.479	1.71	1.308	4.135	2.21	1.487	4.701
1.22	1.105	3.493	1.72	1.311	4.147	2.22	1.490	4.712
1.23	1.109	3.507	1.73	1.315	4.159	2.23	1.493	4.722
1.24	1.114	3.521	1.74	1.319	4.171	2.24	1.497	4.733
1.25	1.118	3.536	1.75	1.323	4.183	2.25	1.500	4.743
1.26	1.122	3.550	1.76	1.327	4.195	2.26	1.503	4.754
1.27	1.127	3.564	1.77	1.330	4.207	2.27	1.507	4.764
1.28	1.131	3.578	1.78	1.334	4.219	2.28	1.510	4.775
1.29	1.136	3.592	1.79	1.338	4.231	2.29	1.513	4.785
1.30	1.140	3.606	1.80	1.342	4.243	2.30	1.517	4.796
1.31	1.145	3.619	1.81	1.345	4.254	2.31	1.520	4.806
1.32	1.149	3.633	1.82	1.349	4.266	2.32	1.523	4.817
1.33	1.153	3.647	1.83	1.353	4.278	2.33	1.526	4.827
1.34	1.158	3.661	1.84	1.356	4.290	2.34	1.530	4.837
1.35	1.162	3.674	1.85	1.360	4.301	2.35	1.533	4.848
1.36	1.166	3.688	1.86	1.364	4.313	2.36	1.536	4.858
1.37	1.170	3.701	1.87	1.367	4.324	2.37	1.539	4.868
1.38	1.175	3.715	1.88	1.371	4.336	2.38	1.543	4.879
1.39	1.179	3.728	1.89	1.375	4.347	2.39	1.546	4.889
1.40	1.183	3.742	1.90	1.378	4.359	2.40	1.549	4.899
1.41	1.187	3.755	1.91	1.382	4.370	2.41	1.552	4.909
1.42	1.192	3.768	1.92	1.386	4.382	2.42	1.556	4.919
1.43	1.196	3.782	1.93	1.389	4.393	2.43	1.559	4.930
1.44	1.200	3.795	1.94	1.393	4.405	2.44	1.562	4.940
1.45	1.204	3.808	1.95	1.396	4.416	2.45	1.565	4.950
1.46	1.208	3.821	1.96	1.400	4.427	2.46	1.568	4.960
1.47	1.212	3.834	1.97	1.404	4.438	2.47	1.572	4.970
1.48	1.217	3.847	1.98	1.407	4.450	2.48	1.575	4.980
1.49	1.221	3.860	1.99	1.411	4.461	2.49	1.578	4.990

N	$\sqrt{N}$	$\sqrt{10N}$	N	$\sqrt{N}$	$\sqrt{10N}$	N	$\sqrt{N}$	$\sqrt{10N}$
2.50	1.581	5.000	3.00	1.732	5.477	3.50	1.871	5.916
2.51	1.584	5.010	3.01	1.735	5.486	3.51	1.873	5.925
2.52	1.587	5.020	3.02	1.738	5.495	3.52	1.876	5.933
2.53	1.591	5.030	3.03	1.741	5.505	3.53	1.879	5.941
2.54	1.594	5.040	3.04	1.744	5.514	3.54	1.881	5.950
2.55	1.597	5.050	3.05	1.746	5.523	3.55	1.884	5.958
2.56	1.600	5.060	3.06	1.749	5.532	3.56	1.887	5.967
2.57	1.603	5.070	3.07	1.752	5.541	3.57	1.889	5.975
2.58	1.606	5.079	3.08	1.755	5.550	3.58	1.892	5.983
2.59	1.609	5.089	3.09	1.758	5.559	3.59	1.895	5.992
2.60	1.612	5.099	3.10	1.761	5.568	3.60	1.897	6.000
2.61	1.616	5.109	3.11	1.764	5.577	3.61	1.900	6.008
2.62	1.619	5.119	3.12	1.766	5.586	3.62	1.903	6.017
2.63	1.622	5.128	3.13	1.769	5.595	3.63	1.905	6.025
2.64	1.625	5.138	3.14	1.772	5.604	3.64	1.908	6.033
2.65	1.628	5.148	3.15	1.775	5.612	3.65	1.910	6.042
2.66	1.631	5.158	3.16	1.778	5.621	3.66	1.913	6.050
2.67	1.634	5.167	3.17	1.780	5.630	3.67	1.916	6.058
2.68	1.637	5.177	3.18	1.783	5.639	3.68	1.918	6.066
2.69	1.640	5.187	3.19	1.786	5.648	3.69	1.921	6.075
2.70	1.643	5.196	3.20	1.789	5.657	3.70	1.924	6.083
2.71	1.646	5.206	3.21	1.792	5.666	3.71	1.926	6.091
2.72	1.649	5.215	3.22	1.794	5.675	3.72	1.929	6.099
2.73	1.652	5.225	3.23	1.797	5.683	3.73	1.931	6.107
2.74	1.655	5.234	3.24	1.800	5.692	3.74	1.934	6.116
2.75	1.658	5.244	3.25	1.803	5.701	3.75	1.936	6.124
2.76	1.661	5.254	3.26	1.806	5.710	3.76	1.939	6.132
2.77	1.664	5.263	3.27	1.808	5.718	3.77	1.942	6.140
2.78	1.667	5.273	3.28	1.811	5.727	3.78	1.944	6.148
2.79	1.670	5.282	3.29	1.814	5.736	3.79	1.947	6.156
2.80	1.673	5.292	3.30	1.817	5.745	3.80	1.949	6.164
2.81	1.676	5.301	3.31	1.819	5.753	3.81	1.852	6.173
2.82	1.679	5.310	3.32	1.822	5.762	3.82	1.954	6.181
2.83	1.682	5.320	3.33	1.825	5.771	3.83	1.957	6.189
2.84	1.685	5.329	3.34	1.828	5.779	3.84	1.960	6.197
2.85	1.688	5.339	3.35	1.830	5.788	3.85	1.962	6.205
2.86	1.691	5.348	3.36	1.833	5.797	3.86	1.965	6.213
2.87	1.694	5.357	3.37	1.836	5.805	3.87	1.967	6.221
2.88	1.697	5.367	3.38	1.838	5.814	3.88	1.970	6.229
2.89	1.700	5.376	3.39	1.841	5.822	3.89	1.972	6.237
2.90	1.703	5.385	3.40	1.844	5.831	3.90	1.975	6.245
2.91	1.706	5.394	3.41	1.847	5.840	3.91	1.977	6.253
2.92	1.709	5.404	3.42	1.849	5.848	3.92	1.980	6.261
2.93	1.712	5.413	3.43	1.852	5.857	3.93	1.982	6.269
2.94	1.715	5.422	3.44	1.855	5.865	3.94	1.985	6.277
2.95	1.718	5.431	3.45	1.857	5.874	3.95	1.987	6.285
2.96	1.720	5.441	3.46	1.860	5.882	3.96	1.990	6.293
2.97	1.723	5.450	3.47	1.863	5.891	3.97	1.992	6.301
2.98	1.726	5.459	3.48	1.865	5.899	3.98	1.995	6.309
2.99	1.729	5.468	3.49	1.868	5.908	3.99	1.997	6.317

N	√N	√10N	N	√N	√10N	N	√N	√10N
4.00	2.000	6.325	4.50	2.121	6.708	5.00	2.236	7.071
4.01	2.002	6.332	4.51	2.124	6.716	5.01	2.238	7.078
4.02	2.005	6.340	4.52	2.126	6.723	5.02	2.241	7.085
4.03	2.007	6.348	4.53	2.128	6.731	5.03	2.243	7.092
4.04	2.010	6.356	4.54	2.131	6.738	5.04	2.245	7.099
4.05	2.012	6.364	4.55	2.133	6.745	5.05	2.247	7.106
4.06	2.015	6.372	4.56	2.135	6.753	5.06	2.249	7.113
4.07	2.017	6.380	4.57	2.138	6.760	5.07	2.252	7.120
4.08	2.020	6.387	4.58	2.140	6.768	5.08	2.254	7.127
4.09	2.022	6.395	4.59	2.142	6.775	5.09	2.256	7.134
4.10	2.025	6.403	4.60	2.145	6.782	5.10	2.258	7.141
4.11	2.027	6.411	4.61	2.147	6.790	5.11	2.261	7.148
4.12	2.030	6.419	4.62	2.149	6.797	5.12	2.263	7.155
4.13	2.032	6.427	4.63	2.152	6.804	5.13	2.265	7.162
4.14	2.035	6.434	4.64	2.154	6.812	5.14	2.267	7.169
4.15	2.037	6.442	4.65	2.156	6.819	5.15	2.269	7.176
4.16	2.040	6.450	4.66	2.159	6.826	5.16	2.272	7.183
4.17	2.042	6.458	4.67	2.161	6.834	5.17	2.274	7.190
4.18	2.045	6.465	4.68	2.163	6.841	5.18	2.276	7.197
4.19	2.047	6.473	4.69	2.166	6.848	5.19	2.278	7.204
4.20	2.049	6.481	4.70	2.168	6.856	5.20	2.280	7.211
4.21	2.052	6.488	4.71	2.170	6.863	5.21	2.280	7.218
4.22	2.054	6.496	4.72	2.173	6.870	5.22	2.285	7.225
4.23	2.057	6.504	4.73	2.175	6.877	5.23	2.287	7.232
4.24	2.059	6.512	4.74	2.177	6.885	5.24	2.289	7.239
4.25	2.062	6.519	4.75	2.179	6.892	5.25	2.291	7.246
4.26	2.064	6.527	4.76	2.182	6.899	5.26	2.293	7.253
4.27	2.066	6.535	4.77	2.184	6.907	5.27	2.296	7.259
4.28	2.069	6.542	4.78	2.186	6.914	5.28	2.298	7.266
4.29	2.071	6.550	4.79	2.189	6.921	5.29	2.300	7.273
4.30	2.074	6.557	4.80	2.191	6.928	5.30	2.302	7.280
4.31	2.076	6.565	4.81	2.193	6.935	5.31	2.304	7.287
4.32	2.078	6.573	4.82	2.195	6.943	5.32	2.307	7.294
4.33	2.081	6.580	4.83	2.198	6.950	5.33	2.309	7.301
4.34	2.083	6.588	4.84	2.200	6.957	5.34	2.311	7.308
4.35	2.086	6.595	4.85	2.202	6.964	5.35	2.313	7.314
4.36	2.088	6.603	4.86	2.205	6.971	5.36	2.315	7.321
4.37	2.090	6.611	4.87	2.207	6.979	5.37	2.317	7.328
4.38	2.093	6.618	4.88	2.209	6.986	5.38	2.319	7.335
4.39	2.095	6.626	4.89	2.211	6.993	5.39	2.322	7.342
4.40	2.098	6.633	4.90	2.214	7.000	5.40	2.324	7.348
4.41	2.100	6.641	4.91	2.216	7.007	5.41	2.326	7.355
4.42	2.102	6.648	4.92	2.218	7.014	5.42	2.328	7.632
4.43	2.105	6.656	4.93	2.220	7.021	5.43	2.330	7.369
4.44	2.107	6.663	4.94	2.223	7.029	5.44	2.332	7.376
4.45	2.110	6.671	4.95	2.225	7.036	5.45	2.335	7.382
4.46	2.112	6.678	4.96	2.227	7.043	5.46	2.337	7.389
4.47	2.114	6.686	4.97	2.229	7.050	5.47	2.339	7.396
4.48	2.117	6.693	4.98	2.232	7.057	5.48	2.341	7.403
4.49	2.119	6.701	4.99	2.234	7.064	5.49	2.343	7.409

N	$\sqrt{N}$	$\sqrt{10N}$	N	$\sqrt{N}$	$\sqrt{10N}$	N	$\sqrt{N}$	$\sqrt{10N}$
5.50	2.345	7.416	6.00	2.449	7.746	6.50	2.550	8.062
5.51	2.347	7.423	6.01	2.452	7.752	6.51	2.551	8.068
5.52	2.349	7.430	6.02	2.454	7.759	6.52	2.553	8.075
5.53	2.352	7.436	6.03	2.456	7.765	6.53	2.555	8.081
5.54	2.354	7.443	6.04	2.458	7.772	6.54	2.557	8.087
5.55	2.356	7.450	6.05	2.460	7.778	6.55	2.559	8.093
5.56	2.358	7.457	6.06	2.462	7.785	6.56	2.561	8.099
5.57	2.360	7.463	6.07	2.464	7.791	6.57	2.563	8.106
5.58	2.362	7.470	6.08	2.466	7.797	6.58	2.565	8.112
5.59	2.364	7.477	6.09	2.468	7.804	6.59	2.567	8.118
5.60	2.366	7.483	6.10	2.470	7.810	6.60	2.569	8.124
5.61	2.369	7.490	6.11	2.472	7.817	6.61	2.571	8.130
5.62	2.371	7.497	6.12	2.474	7.823	6.62	2.573	8.136
5.63	2.373	7.503	6.13	2.476	7.829	6.63	2.575	8.142
5.64	2.375	7.510	6.14	2.478	7.836	6.64	2.577	8.149
5.65	2.377	7.517	6.15	2.480	7.842	6.65	2.579	8.155
5.66	2.379	7.523	6.16	2.482	7.849	6.66	2.581	8.161
5.67	2.381	7.530	6.17	2.484	7.855	6.67	2.583	8.167
5.68	2.383	7.537	6.18	2.486	7.861	6.68	2.585	8.173
5.69	2.385	7.543	6.19	2.488	7.868	6.69	2.587	8.179
5.70	2.387	7.550	6.20	2.490	7.874	6.70	2.588	8.185
5.71	2.390	7.556	6.21	2.492	7.880	6.71	2.590	8.191
5.72	2.392	7.563	6.22	2.494	7.887	6.72	2.592	8.198
5.73	2.394	7.570	6.23	2.496	7.893	6.73	2.594	8.204
5.74	2.396	7.576	6.24	2.498	7.899	6.74	2.596	8.210
5.75	2.398	7.583	6.25	2.500	7.906	6.75	2.598	8.216
5.76	2.400	7.589	6.26	2.502	7.912	6.76	2.600	8.222
5.77	2.402	7.596	6.27	2.504	7.918	6.77	2.602	8.228
5.78	2.404	7.603	6.28	2.506	7.925	6.78	2.604	8.234
5.79	2.406	7.609	6.29	2.508	7.931	6.79	2.606	8.240
5.80	2.408	7.616	6.30	2.510	7.937	6.80	2.608	8.246
5.81	2.410	7.622	6.31	2.512	7.944	6.81	2.610	8.252
5.82	2.412	7.629	6.32	2.514	7.950	6.82	2.612	8.258
5.83	2.415	7.635	6.33	2.516	7.956	6.83	2.613	8.264
5.84	2.417	7.642	6.34	2.518	7.962	6.84	2.615	8.270
5.85	2.419	7.649	6.35	2.520	7.969	6.85	2.617	8.276
5.86	2.421	7.655	6.36	2.522	7.975	6.86	2.619	8.283
5.87	2.423	7.662	6.37	2.524	7.981	6.87	2.621	8.289
5.88	2.425	7.668	6.38	2.526	7.987	6.88	2.623	8.295
5.89	2.427	7.675	6.39	2.528	7.994	6.89	2.625	8.301
5.90	2.429	7.681	6.40	2.530	8.000	6.90	2.627	8.307
5.91	2.431	7.688	6.41	2.532	8.006	6.91	2.629	8.313
5.92	2.433	7.694	6.42	2.534	8.012	6.92	2.631	8.319
5.93	2.435	7.701	6.43	2.536	8.019	6.93	2.632	8.325
5.94	2.437	7.707	6.44	2.538	8.025	6.94	2.634	8.331
5.95	2.439	7.714	6.45	2.540	8.031	6.95	2.636	8.337
5.96	2.441	7.720	6.46	2.542	8.037	6.96	2.638	8.343
5.97	2.443	7.727	6.47	2.544	8.044	6.97	2.640	8.349
5.98	2.445	7.733	6.48	2.546	8.50	6.98	2.642	8.355
5.99	2.447	7.740	6.49	2.548	8.056	6.99	2.644	8.361

N	$\sqrt{N}$	$\sqrt{10N}$	N	$\sqrt{N}$	$\sqrt{10N}$	N	$\sqrt{N}$	$\sqrt{10N}$
7.00	2.646	8.367	7.50	2.739	8.660	8.00	2.828	8.944
7.01	2.648	8.373	7.51	2.740	8.666	8.01	2.830	8.950
7.02	2.650	8.379	7.52	2.742	8.672	8.02	2.832	8.955
7.03	2.651	8.385	7.53	2.744	8.678	8.03	2.834	8.961
7.04	2.653	8.390	7.54	2.746	8.683	8.04	2.835	8.967
7.05	2.655	8.396	7.55	2.748	8.689	8.05	2.837	8.972
7.06	2.657	8.402	7.56	2.750	8.695	8.06	2.839	8.978
7.07	2.659	8.408	7.57	2.751	8.701	8.07	2.841	8.983
7.08	2.661	8.414	7.58	2.753	8.706	8.08	2.843	8.989
7.09	2.663	8.420	7.59	2.755	8.712	8.09	2.844	8.994
7.10	2.665	8.426	7.60	2.757	8.718	8.10	2.846	9.000
7.11	2.666	8.432	7.61	2.759	8.724	8.11	2.848	9.006
7.12	2.668	8.438	7.62	2.760	8.729	8.12	2.850	9.011
7.13	2.670	8.444	7.63	2.762	8.735	8.13	2.851	9.017
7.14	2.672	8.450	7.64	2.764	8.741	8.14	2.853	9.022
7.15	2.674	8.456	7.65	2.766	8.746	8.15	2.855	9.028
7.16	2.676	8.462	7.66	2.768	8.752	8.16	2.857	9.033
7.17	2.678	8.468	7.67	2.769	8.758	8.17	2.858	9.039
7.18	2.680	8.473	7.68	2.771	8.764	8.18	2.860	9.044
7.19	2.681	8.479	7.69	2.773	8.769	8.19	2.862	9.050
7.20	2.683	8.485	7.70	2.775	8.775	8.20	2.864	9.055
7.21	2.685	8.491	7.71	2.777	8.781	8.21	2.865	9.061
7.22	2.687	8.497	7.72	2.778	8.786	8.22	2.867	9.066
7.23	2.689	8.503	7.73	2.780	8.792	8.23	2.869	9.072
7.24	2.691	8.509	7.74	2.782	8.798	8.24	2.871	9.077
7.25	2.693	8.515	7.75	2.784	8.803	8.25	2.872	9.083
7.26	2.694	8.521	7.76	2.786	8.809	8.26	2.874	9.088
7.27	2.696	8.526	7.77	2.787	8.815	8.27	2.876	9.094
7.28	2.698	8.532	7.78	2.789	8.820	8.28	2.877	9.099
7.29	2.700	8.538	7.79	2.791	8.826	8.29	2.879	9.105
7.30	2.702	8.544	7.80	2.793	8.832	8.30	2.881	9.110
7.31	2.704	8.550	7.81	2.795	8.837	8.31	2.883	9.116
7.32	2.706	8.556	7.82	2.796	8.843	8.32	2.884	9.121
7.33	2.707	8.562	7.83	2.798	8.849	8.33	2.886	9.127
7.34	2.709	8.567	7.84	2.800	8.854	8.34	2.888	9.132
7.35	2.711	8.573	7.85	2.802	8.860	8.35	2.890	9.138
7.36	2.713	8.579	7.86	2.804	8.866	8.36	2.891	9.143
7.37	2.715	8.585	7.87	2.805	8.871	8.37	2.893	9.149
7.38	2.717	8.591	7.88	2.807	8.877	8.38	2.895	9.154
7.39	2.718	8.597	7.89	2.809	8.883	8.39	2.897	9.160
7.40	2.720	8.602	7.90	2.811	8.888	8.40	2.898	9.165
7.41	2.722	8.608	7.91	2.812	8.894	8.41	2.900	9.171
7.42	2.724	8.614	7.92	2.814	8.899	8.42	2.902	9.176
7.43	2.726	8.620	7.93	2.816	8.905	8.43	2.903	9.182
7.44	2.728	8.626	7.94	2.818	8.911	8.44	2.905	9.187
7.45	2.729	8.631	7.95	2.820	8.916	8.45	2.907	9.192
7.46	2.731	8.637	7.96	2.821	8.922	8.46	2.909	9.198
7.47	2.733	8.643	7.97	2.823	8.927	8.47	2.910	9.203
7.48	2.735	8.649	7.98	2.825	8.933	8.48	2.912	9.209
7.49	2.737	8.654	7.99	2.827	8.939	8.49	2.914	9.214

N	$\sqrt{N}$	$\sqrt{10N}$	N	$\sqrt{N}$	$\sqrt{10N}$	N	$\sqrt{N}$	$\sqrt{10N}$
8.50	2.915	9.220	9.00	3.000	9.480	9.50	3.082	9.747
8.51	2.917	9.225	9.01	3.002	9.492	9.51	3.084	9.752
8.52	2.919	9.230	9.02	3.003	9.497	9.52	3.085	9.757
8.53	2.921	9.236	9.03	3.005	9.503	9.53	3.087	9.762
8.54	2.922	9.241	9.04	3.007	9.508	9.54	3.089	9.767
8.55	2.924	9.247	9.05	3.008	9.513	9.55	3.090	9.772
8.56	2.926	9.252	9.06	3.010	9.518	9.56	3.092	9.778
8.57	2.927	9.257	9.07	3.012	9.524	9.57	3.094	9.783
8.58	2.929	9.263	9.08	3.013	9.529	9.58	3.095	9.788
8.59	2.931	9.268	9.09	3.015	9.534	9.59	3.097	9.793
8.60	2.933	9.274	9.10	3.017	9.539	9.60	3.098	9.798
8.61	2.934	9.279	9.11	3.017	9.545	9.61	3.100	9.803
8.62	2.936	9.284	9.12	3.020	9.550	9.62	3.102	9.808
8.63	2.938	9.290	9.13	3.022	9.555	9.63	3.103	9.813
8.64	2.939	9.295	9.14	3.023	9.560	9.64	3.105	9.818
8.65	2.941	9.301	9.15	3.025	9.566	9.65	3.106	9.823
8.66	2.943	9.306	9.16	3.027	9.571	9.66	3.108	9.829
8.67	2.944	9.311	9.17	3.028	9.576	9.67	3.110	9.834
8.68	2.946	9.317	9.18	3.030	9.581	9.68	3.111	9.839
8.69	2.948	9.322	9.19	3.031	9.586	9.69	3.113	9.844
8.70	2.950	9.327	9.20	3.033	9.592	9.70	3.114	9.849
8.71	2.951	9.333	9.21	3.035	9.597	9.71	3.116	9.854
8.72	2.953	9.338	9.22	3.036	9.602	9.72	3.118	9.859
8.73	2.955	9.343	9.23	3.038	9.607	9.73	3.119	9.864
8.74	2.956	9.349	9.24	3.040	9.612	9.74	3.121	9.869
8.75	2.958	9.354	9.25	3.041	9.618	9.75	3.122	9.874
8.76	2.960	9.359	9.26	3.043	9.623	9.76	3.124	9.879
8.77	2.961	9.365	9.27	3.045	9.628	9.77	3.126	9.884
8.78	2.963	9.370	9.28	3.046	9.633	9.78	3.127	9.889
8.79	2.965	9.375	9.29	3.048	9.638	9.79	3.129	9.894
8.80	2.966	9.381	9.30	3.050	9.644	9.80	3.130	9.899
8.81	2.968	9.386	9.31	3.051	9.649	9.81	3.132	9.905
8.82	2.970	9.391	9.32	3.053	9.654	9.82	3.134	9.910
8.83	2.972	9.397	9.33	3.055	9.659	9.83	3.135	9.915
8.84	2.973	9.402	9.34	3.056	9.664	9.84	3.137	9.920
8.85	2.975	9.407	9.35	3.058	9.670	9.85	3.138	9.925
8.86	2.977	9.413	9.36	3.059	9.675	9.86	3.140	9.930
8.87	2.978	9.418	9.37	3.061	9.680	9.87	3.142	9.935
8.88	2.980	9.423	9.38	3.063	9.685	9.88	3.143	9.940
8.89	2.982	9.429	9.39	3.064	9.690	9.89	3.145	9.945
8.90	2.983	9.434	9.40	3.066	9.695	9.90	3.146	9.950
8.91	2.985	9.439	9.41	3.068	9.701	9.91	3.148	9.955
8.92	2.987	9.445	9.42	3.069	9.706	9.92	3.150	9.960
8.93	2.988	9.450	9.43	3.071	9.711	9.93	3.151	9.965
8.94	2.990	9.455	9.44	3.072	9.716	9.94	3.153	9.970
8.95	2.992	9.460	9.45	3.074	9.721	9.95	3.154	9.975
8.96	2.993	9.466	9.46	3.076	9.726	9.96	3.156	9.980
8.97	2.995	9.471	9.47	3.077	9.731	9.97	3.158	9.985
8.98	2.997	9.476	9.48	3.079	9.737	9.98	3.158	9.990
8.99	2.998	9.482	9.49	3.081	9.742	9.99	3.161	9.995

SOURCE: M. Orkin and R. Drogin, "Vital Statistics," pp. 332–337, McGraw-Hill, New York, 1975.

TABLE A.2
Binomial distribution

Table A.2 gives cumulative binomial probabilities $P(S \le k)$ for a variable S having a binomial distribution with parameters n and p. For example, if $n = 10$ and $p = .40$, then $P(S \le 3) = .3823$, or if $n = 13$ and $p = .85$, then $P(S \le 10) = .2704$. Other probabilities of interest may be obtained according to the formulas

$$P(S > k) = 1 - P(S \le k)$$

$$P(S = k) = P(S \le k) - P(S \le k - 1)$$

n	k	$p = .05$	.10	.15	.20	.25	.30	.35	.40	.45
1	0	.9500	.9000	.8500	.8000	.7500	.7000	.6500	.6000	.5500
	1	1.0000	1.0000	1.0000	1.0000	1.0000	1.0000	1.0000	1.0000	1.0000
2	0	.9025	.8100	.7225	.6400	.5625	.4900	.4225	.3600	.3025
	1	.9975	.9900	.9775	.9600	.9375	.9100	.8775	.8400	.7975
	2	1.0000	1.0000	1.0000	1.0000	1.0000	1.0000	1.0000	1.0000	1.0000
3	0	.8574	.7290	.6141	.5129	.4219	.3439	.2746	.2160	.1664
	1	.9928	.9720	.9392	.8960	.8438	.7840	.7182	.6480	.5748
	2	.9999	.9990	.9966	.9929	.9844	.9730	.9571	.9360	.9089
	3	1.0000	1.0000	1.0000	1.0000	1.0000	1.0000	1.0000	1.0000	1.0000
4	0	.8145	.6561	.5200	.4096	.3164	.2401	.1785	.1296	.0915
	1	.9860	.9477	.8905	.8192	.7383	.6517	.5630	.4752	.3910
	2	.9995	.9963	.9880	.9728	.9492	.9163	.8735	.8208	.7585
	3	1.0000	.9999	.9995	.9984	.9961	.9919	.9850	.9743	.9590
	4	1.0000	1.0000	1.0000	1.0000	1.0000	1.0000	1.0000	1.0000	1.0000
5	0	.7738	.5905	.4437	.3277	.2373	.1681	.1160	.0778	.0503
	1	.9774	.9185	.8352	.7373	.6328	.5282	.4284	.3370	.2562
	2	.9988	.9924	.9734	.9421	.8965	.8369	.7648	.6826	.5931
	3	1.0000	.9995	.9978	.9933	.9844	.9692	.9460	.9130	.8688
	4	1.0000	1.0000	.9999	.9997	.9990	.9976	.9947	.9898	.9815
	5	1.0000	1.0000	1.0000	1.0000	1.0000	1.0000	1.0000	1.0000	1.0000
6	0	.7351	.5314	.3771	.2621	.1780	.1176	.0754	.0467	.0277
	1	.9672	.8857	.7765	.6534	.5339	.4202	.3191	.2333	.1636
	2	.9978	.9842	.9527	.9011	.8306	.7443	.6471	.5443	.4415
	3	.9999	.9987	.9941	.9830	.9624	.9295	.8826	.8208	.7447
	4	1.0000	.9999	.9996	.9984	.9954	.9891	.9777	.9590	.9308
	5	1.0000	1.0000	1.0000	.9999	.9998	.9993	.9982	.9959	.9917
	6	1.0000	1.0000	1.0000	1.0000	1.0000	1.0000	1.0000	1.0000	1.0000
7	0	.6983	.4783	.3206	.2097	.1335	.0824	.0490	.0280	.0152
	1	.9556	.8503	.7166	.5767	.4449	.3294	.2338	.1586	.1024
	2	.9962	.9743	.9262	.8520	.7564	.6471	.5323	.4199	.3164
	3	.9998	.9973	.9879	.9667	.9294	.8740	.8002	.7102	.6083
	4	1.0000	.9998	.9988	.9953	.9871	.9712	.9444	.9037	.8643
	5	1.0000	1.0000	.9999	.9996	.9987	.9962	.9910	.9812	.9643
	6	1.0000	1.0000	1.0000	1.0000	.9999	.9998	.9994	.9984	.0063
	7	1.0000	1.0000	1.0000	1.0000	1.0000	1.0000	1.0000	1.0000	1.0000

p = .50	.55	.60	.65	.70	.75	.80	.85	.90	.95
.5000	.4500	.4000	.3500	.3000	.2500	.2000	.1500	.1000	.0500
1.0000	1.0000	1.0000	1.0000	1.0000	1.0000	1.0000	1.0000	1.0000	1.0000
.2500	.2025	.1600	.1225	.0900	.0625	.0400	.0225	.0100	.0025
.7500	.6975	.6400	.5775	.5100	.4373	.3600	.2775	.1900	.0975
1.0000	1.0000	1.0000	1.0000	1.0000	1.0000	1.0000	1.0000	1.0000	1.0000
.1250	.0911	.0640	.0429	.0270	.0156	.0080	.0034	.0010	.0001
.5000	.4252	.3520	.2818	.2160	.1562	.1040	.0608	.0280	.0072
.8750	.8336	.7840	.7254	.6570	.5781	.4880	.3959	.2710	.1426
1.0000	1.0000	1.0000	1.0000	1.0000	1.0000	1.0000	1.0000	1.0000	1.0000
.0625	.0410	.0256	.0150	.0081	.0039	.0016	.0005	.0001	.0000
.3125	.2415	.1792	.1265	.0837	.0508	.0272	.0120	.0037	.0005
.6875	.6090	.5248	.4370	.3483	.2617	.1808	.1095	.0523	.0140
.9375	.9085	.8704	.8215	.7599	.6836	.5904	.4780	.3439	.1855
1.0000	1.0000	1.0000	1.0000	1.0000	1.0000	1.0000	1.0000	1.0000	1.0000
.0312	.0185	.0102	.0053	.0024	.0010	.0003	.0001	.0000	.0000
.1875	.1312	.0870	.0540	.0308	.0156	.0067	.0022	.0005	.0000
.5000	.4069	.3174	.2352	.1631	.1035	.0579	.0266	.0086	.0012
.8125	.7438	.6630	.5716	.4718	.3672	.2627	.1648	.0815	.0226
.9688	.9497	.9222	.8840	.8319	.7627	.6723	.5563	.4095	.2262
1.0000	1.0000	1.0000	1.0000	1.0000	1.0000	1.0000	1.0000	1.0000	1.0000
.0156	.0083	.0041	.0018	.0007	.0002	.0001	.0000	.0000	.0000
.1094	.0692	.0410	.0023	.0109	.0046	.0016	.0004	.0001	.0000
.3438	.2553	.1792	.1174	.0705	.0376	.0170	.0059	.0013	.0001
.6562	.5585	.4557	.3529	.2557	.1694	.0989	.0473	.0158	.0022
.8906	.8364	.7667	.6809	.5798	.4661	.3446	.2235	.1143	.0328
.9844	.9723	.9533	.9246	.8824	.8220	.7379	.6229	.4686	.2649
1.0000	1.0000	1.0000	1.0000	1.0000	1.0000	1.0000	1.0000	1.0000	1.0000
.0078	.0037	.0016	.0006	.0002	.0001	.0000	.0000	.0000	.0000
.0625	.0357	.0188	.0090	.0038	.0013	.0004	.0001	.0000	.0000
.2266	.1529	.0963	.0556	.0288	.0129	.0047	.0012	.0002	.0000
.5000	.3917	.2898	.1998	.1260	.0706	.0333	.0121	.0027	.0002
.7734	.6836	.5801	.4677	.3529	.2436	.1480	.0738	.0257	.0038
.9375	.8976	.8414	.7662	.6706	.5551	.4233	.2834	.1497	.0444
.9922	.9848	.9720	.9510	.9176	.8665	.7903	.6794	.5217	.3917
1.0000	1.0000	1.0000	1.0000	1.0000	1.0000	1.0000	1.0000	1.0000	1.0000

TABLE A.2 (Continued)

n	k	p = .05	.10	.15	.20	.25	.30	.35	.40	.45
8	0	.6634	.4305	.2725	.1678	.1001	.0576	.0319	.0168	.0084
	1	.9428	.8131	.6572	.5033	.3671	.2553	.1691	.1064	.0632
	2	.9942	.9619	.8948	.7969	.6785	.5518	.4278	.3154	.2201
	3	.9996	.9950	.9786	.9437	.8862	.8059	.7064	.5941	.4770
	4	1.0000	.9996	.9971	.9896	.9727	.9420	.8939	.8263	.7396
	5	1.0000	1.0000	.9998	.9988	.9958	.9887	.9747	.9502	.9115
	6	1.0000	1.0000	1.0000	.9999	.9996	.9987	.9964	.9915	.9819
	7	1.0000	1.0000	1.0000	1.0000	1.0000	.9999	.9988	.9993	.9983
	8	1.0000	1.0000	1.0000	1.0000	1.0000	1.0000	1.0000	1.0000	1.0000
9	0	.6302	.3874	.2316	.1342	.0751	.0404	.0207	.0101	.0046
	1	.9288	.7748	.5995	.4362	.3003	.1960	.1211	.0705	.0385
	2	.9916	.9470	.8591	.7382	.6007	.4628	.3373	.2318	.1495
	3	.9994	.9917	.9661	.9144	.8343	.7297	.6089	.4826	.3614
	4	1.0000	.9991	.9944	.9804	.9511	.9012	.8283	.7334	.6214
	5	1.0000	.9999	.9994	.9969	.9900	.9747	.9464	.9006	.9342
	6	1.0000	1.0000	1.0000	.9997	.9987	.9957	.9888	.9750	.9502
	7	1.0000	1.0000	1.0000	1.0000	.9999	.9996	.9986	.9962	.9909
	8	1.0000	1.0000	1.0000	1.0000	1.0000	1.0000	.9999	.9997	.9992
	9	1.0000	1.0000	1.0000	1.0000	1.0000	1.0000	1.0000	1.0000	1.0000
10	0	.5987	.3487	.1969	.1074	.0563	.0282	.0135	.0060	.0025
	1	.9139	.7361	.5443	.3758	.2440	.1493	.0860	.0464	.0233
	2	.9885	.9298	.8202	.6778	.5256	.3828	.2616	.1673	.0996
	3	.9990	.9872	.9500	.8791	.7759	.6496	.5138	.3823	.2660
	4	.9999	.9984	.9901	.9672	.9219	.8497	.7515	.6331	.5044
	5	1.0000	.9999	.9986	.9936	.9803	.9527	.9051	.8338	.7384
	6	1.0000	1.0000	.9999	.9991	.9965	.9894	.9740	.9452	.8980
	7	1.0000	1.0000	1.0000	.9999	.9996	.9984	.9952	.9877	.9726
	8	1.0000	1.0000	1.0000	1.0000	1.0000	.9999	.9995	.9983	.9955
	9	1.0000	1.0000	1.0000	1.0000	1.0000	1.0000	1.0000	.9999	.9997
	10	1.0000	1.0000	1.0000	1.0000	1.0000	1.0000	1.0000	1.0000	1.0000
11	0	.5688	.3138	.1673	.0859	.0422	.0198	.0088	.0036	.0014
	1	.8981	.6974	.4922	.3221	.1971	.1130	.0606	.0302	.0139
	2	.9848	.9104	.7788	.6174	.4552	.3127	.2001	.1189	.0652
	3	.9984	.9815	.9306	.8389	.7133	.5696	.4256	.2963	.1911
	4	.9999	.9972	.9841	.9496	.8854	.7897	.6683	.5328	.3971
	5	1.0000	.9997	.9973	.9883	.9657	.9218	.8513	.7535	.6331
	6	1.0000	1.0000	.9997	.9980	.9924	.9784	.9499	.9006	.8262
	7	1.0000	1.0000	1.0000	.9998	.9988	.9957	.9878	.9707	.9390
	8	1.0000	1.0000	1.0000	1.0000	.9999	.9994	.9980	.9941	.9852
	9	1.0000	1.0000	1.0000	1.0000	1.0000	1.0000	.9998	.9993	.9978
	10	1.0000	1.0000	1.0000	1.0000	1.0000	1.0000	1.0000	1.0000	.9998
	11	1.0000	1.0000	1.0000	1.0000	1.0000	1.0000	1.0000	1.0000	1.0000

p = .50	.55	.60	.65	.70	.75	.80	.85	.90	.95
.0030	.0017	.0007	.9992	.0001	.0000	.0000	.0000	.0000	.0000
.0352	.0181	.0085	.0086	.0013	.0004	.0001	.0000	.0000	.0000
.1445	.0885	.0498	.0253	.0113	.0042	.0012	.0002	.0000	.0000
.3633	.2604	.1737	.1061	.0580	.0273	.0104	.0029	.0004	.0000
.6367	.5230	.4059	.2936	.1941	.1138	.0563	.0214	.0050	.0004
.8555	.7799	.6846	.5722	.4482	.3215	.2031	.1052	.0381	.0058
.9648	.9368	.8936	.8309	.7447	.6329	.4967	.3428	.1869	.0572
.9961	.9916	.9832	.9681	.9424	.8999	.8322	.7275	.5695	.3366
1.0000	1.0000	1.0000	1.0000	1.0000	1.0000	1.0000	1.0000	1.0000	1.0000
.0020	.0008	.0003	.0001	.0000	.0000	.0000	.0000	.0000	.0000
.0195	.0091	.0038	.0014	.0004	.0001	.0000	.0000	.0000	.0000
.0898	.0498	.0250	.0112	.0043	.0013	.0003	.0000	.0000	.0000
.2539	.1658	.0994	.0536	.0253	.0100	.0031	.0006	.0001	.0000
.5000	.3786	.2666	.1717	.0988	.0489	.0196	.0056	.0009	.0000
.7461	.6386	.5174	.3911	.2703	.1657	.0856	.0339	.0083	.0006
.9102	.8505	.7682	.6627	.5372	.3993	.2618	.1409	.0530	.0084
.9805	.9615	.9295	.8789	.8040	.6997	.5638	.4005	.2252	.0712
.9980	.9954	.9899	.9793	.9596	.9249	.8658	.7684	.6126	.3698
1.0000	1.0000	1.0000	1.0000	1.0000	1.0000	1.0000	1.0000	1.0000	1.0000
.0010	.0003	.0001	.0000	.0000	.0000	.0000	.0000	.0000	.0000
.0107	.0045	.0017	.0005	.0001	.0000	.0000	.0000	.0000	.0000
.0547	.0274	.0123	.0048	.0016	.0004	.0001	.0000	.0000	.0000
.1719	.1020	.0548	.0260	.0106	.0035	.0009	.0001	.0000	.0000
.3770	.2616	.1662	.0949	.0473	.0197	.0064	.0014	.0001	.0000
.6230	.4956	.3669	.2485	.1503	.0781	.0328	.0099	.0016	.0001
.8281	.7340	.6177	.4862	.3504	.2241	.1209	.0500	.0128	.0010
.9453	.9004	.8327	.7184	.6172	.4744	.3222	.1798	.0702	.0115
.9893	.9767	.9536	.9140	.8507	.7560	.6242	.4557	.2639	.0861
.9990	.9975	.9940	.9865	.9718	.9437	.8926	.8031	.6513	.4013
1.0000	1.0000	1.0000	1.0000	1.0000	1.0000	1.0000	1.0000	1.0000	1.0000
.0005	.0002	.0000	.0000	.0000	.0000	.0000	.0000	.0000	.0000
.0059	.0022	.0007	.0002	.0000	.0000	.0000	.0000	.0000	.0000
.0327	.0148	.0059	.0020	.0006	.0001	.0000	.0000	.0000	.0000
.1133	.0610	.0293	.0122	.0043	.0012	.0002	.0000	.0000	.0000
.2744	.1738	.0994	.0501	.0216	.0076	.0020	.0003	.0000	.0000
.5000	.3669	.2465	.1487	.0782	.0343	.0117	.0027	.0003	.0000
.7256	.6029	.4672	.3317	.2103	.1146	.0504	.0159	.0028	.0001
.8867	.8089	.7037	.5744	.4304	.2867	.1611	.0694	.0185	.0016
.9673	.9348	.8811	.7999	.6873	.5448	.3826	.2212	.0896	.0152
.9941	.9861	.9698	.9394	.8870	.8029	.6779	.5078	.3026	.1019
.9995	.9986	.9964	.9912	.9802	.0578	.9141	.8327	.6862	.4312
1.0000	1.0000	1.0000	1.0000	1.0000	1.0000	1.0000	1.0000	1.0000	1.0000

TABLE A.2 (Continued)

n	k	p = .05	.10	.15	.20	.25	.30	.35	.40	.45
12	0	.5404	.2824	.1422	.0687	.0317	.0138	.0057	.0022	.0008
	1	.8816	.6590	.4435	.2749	.1584	.0850	.0424	.0196	.0083
	2	.9804	.8891	.7358	.5583	.3907	.2528	.1513	.0834	.0421
	3	.9978	.9744	.9078	.7946	.6488	.4925	.3467	.2253	.1345
	4	.9998	.9956	.9761	.9274	.8424	.7237	.5833	.4382	.3044
	5	1.0000	.9995	.9954	.9806	.9456	.8822	.7873	.6652	.5269
	6	1.0000	.9999	.9993	.9961	.9857	.9614	.9154	.8418	.7393
	7	1.0000	1.0000	.9999	.9994	.9972	.9905	.9745	.9427	.8883
	8	1.0000	1.0000	1.0000	.9999	.9996	.9983	.9944	.9847	.9644
	9	1.0000	1.0000	1.0000	1.0000	1.0000	.9998	.9992	.9972	.9921
	10	1.0000	1.0000	1.0000	1.0000	1.0000	1.0000	.9999	.9997	.9989
	11	1.0000	1.0000	1.0000	1.0000	1.0000	1.0000	1.0000	1.0000	.9999
	12	1.0000	1.0000	1.0000	1.0000	1.0000	1.0000	1.0000	1.0000	1.0000
13	0	.5133	.2542	.1209	.0550	.0238	.0097	.0037	.0013	.0004
	1	.8646	.6213	.3983	.2336	.1267	.0637	.9296	.0126	.0049
	2	.9755	.8661	.7296	.5017	.3326	.2025	.1132	.0579	.0269
	3	.9969	.9658	.9033	.7473	.5843	.4206	.2783	.1686	.0929
	4	.9997	.9935	.9740	.9009	.7940	.6543	.5005	.3530	.2279
	5	1.0000	.9991	.9947	.9700	.9198	.8346	.7159	.5744	.4268
	6	1.0000	.9999	.9987	.9930	.9757	.9376	.8705	.7712	.6437
	7	1.0000	1.0000	.9998	.9988	.9944	.9818	.9538	.9023	.8212
	8	1.0000	1.0000	1.0000	.9998	.9990	.9960	.9874	.9679	.9302
	9	1.0000	1.0000	1.0000	1.0000	.9999	.9993	.9975	.9922	.9797
	10	1.0000	1.0000	1.0000	1.0000	1.0000	.9999	.9997	.9987	.9959
	11	1.0000	1.0000	1.0000	1.0000	1.0000	10000	1.0000	.9999	.9995
	12	1.0000	1.0000	1.0000	1.0000	1.0000	1.0000	1.0000	1.0000	1.0000
	13	1.0000	1.0000	1.0000	1.0000	1.0000	1.0000	1.0000	1.0000	1.0000
14	0	.4877	.2288	.1028	.0440	.0178	.0068	.0024	.0008	.0002
	1	.8470	.5846	.3567	.1979	.1010	.0475	.0205	.0081	.0029
	2	.9699	.8416	.6479	.4481	.2811	.1608	.0839	.0398	.0170
	3	.9958	.9559	.8535	.6982	.5213	.3552	.2205	.1243	.0632
	4	.9996	.9908	.9533	.8702	.7415	.5842	.4227	.2793	.1672
	5	1.0000	.9985	.9885	.9561	.8883	.7805	.6405	.4859	.3373
	6	1.0000	.9998	.9978	.9884	.9617	.9067	.8164	.6925	.5461
	7	1.0000	1.0000	.9997	.9976	.9897	.9685	.9247	.8499	.7414
	8	1.0000	1.0000	1.0000	.9996	.9978	.9917	.9757	.9417	.8811
	9	1.0000	1.0000	1.0000	1.0000	.9997	.9983	.9940	.9825	.9574
	10	1.0000	1.0000	1.0000	1.0000	1.0000	.9998	.9989	.9961	.9886
	11	1.0000	1.0000	1.0000	1.0000	1.0000	1.0000	.9999	.9994	.9978
	12	1.0000	1.0000	1.0000	1.0000	1.0000	1.0000	1.0000	.9999	.9997
	13	1.0000	1.0000	1.0000	1.0000	1.0000	1.0000	1.0000	1.0000	1.0000
	14	1.0000	1.0000	1.0000	1.0000	1.0000	1.0000	1.0000	1.0000	1.0000

$p = .50$	.55	.60	.65	.70	.75	.80	.85	.90	.95
.0002	.0001	.0000	.0000	.0000	.0000	.0000	.0000	.0000	.0000
.0032	.0011	.0003	.0001	.0000	.0000	.0000	.0000	.0000	.0000
.0193	.0079	.0028	.0008	.0002	.0000	.0000	.0000	.0000	.0000
.0730	.0356	.0153	.0056	.0017	.0004	.0001	.0000	.0000	.0000
.1938	.1117	.0573	.0255	.0095	.0028	.0006	.0001	.0000	.0000
.3872	.2607	.1582	.0846	.0386	.0143	.0039	.0007	.0001	.0000
.6128	.4731	.3348	.2127	.1178	.0544	.0194	.0046	.0005	.0000
.8062	.6956	.5618	.4167	.2763	.1576	.0726	.0239	.0043	.0002
.9270	.8655	.7747	.6533	.5075	.3512	.2054	.0922	.0256	.0022
.9807	.9579	.9166	.8487	.7472	.6093	.4417	.2642	.1109	.0196
.9968	.9917	.9804	.9576	.9150	.8416	.7251	.5565	.3410	.1184
.9998	.9992	.9978	.9943	.9862	.9683	.9313	.8578	.7176	.4596
1.0000	1.0000	1.0000	1.0000	1.0000	1.0000	1.0000	1.0000	1.0000	1.0000
.0001	.0000	.0000	.0000	.0000	.0000	.0000	.0000	.0000	.0000
.0017	.0005	.0001	.0000	.0000	.0000	.0000	.0000	.0000	.0000
.0112	.0041	.0013	.0003	.0001	.0000	.0000	.0000	.0000	.0000
.0461	.0203	.0078	.0025	.0007	.0001	.0000	.0000	.0000	.0000
.1334	.0698	.0321	.0126	.0040	.0010	.0002	.0000	.0000	.0000
.2905	.1788	.0977	.0462	.0182	.0056	.0012	.0002	.0000	.0000
.5000	.3563	.2288	.1295	.0624	.0243	.0070	.0013	.0001	.0000
.7095	.5732	.4256	.2841	.1654	.0802	.0300	.0053	.0009	.0000
.8666	.7721	.6470	.4995	.3457	.2060	.0991	.0260	.0065	.0003
.9539	.9071	.8314	.7217	.5794	.4157	.2527	.0967	.0342	.0031
.9888	.9731	.9421	.8868	.7975	.6674	.4983	.2704	.1339	.0245
.9983	.9951	.9874	.9704	.9363	.8733	.7664	.6017	.3787	.1354
.9999	.9996	.9987	.9963	.9903	.9762	.9450	.8791	.7458	.4867
1.0000	1.0000	1.0000	1.0000	1.0000	1.0000	1.0000	1.0000	1.0000	1.0000
.0000	.0000	.0000	.0000	.0000	.0000	.0000	.0000	.0000	.0000
.0009	.0003	.0001	.0000	.0000	.0000	.0000	.0000	.0000	.0000
.0065	.0022	.0006	.0001	.0000	.0000	.0000	.0000	.0000	.0000
.0287	.0114	.0039	.0011	.0002	.0000	.0000	.0000	.0000	.0000
.0898	.0462	.0175	.0060	.0017	.0003	.0000	.0000	.0000	.0000
.2120	.1189	.0583	.0243	.0083	.0022	.0004	.0000	.0000	.0000
.3953	.2586	.1501	.0753	.0315	.0108	.0024	.0003	.0000	.0000
.6047	.4539	.3075	.1836	.0933	.0383	.0116	.0022	.0002	.0000
.7880	.6627	.5141	.3595	.2195	.1117	.0439	.0115	.0015	.0000
.9102	.8328	.7207	.5773	.4158	.2585	.1298	.0467	.0092	.0004
.9713	.9368	.8757	.7795	.6448	.4787	.3018	.1465	.0441	.0042
.9935	.9830	.9602	.9161	.8392	.7189	.5519	.3521	.1584	.0301
.9991	.9971	.9919	.9795	.9525	.8990	.8021	.6433	.4154	.1530
.9999	.9998	.9992	.9976	.9932	.9822	.9560	.8972	.7712	.5123
1.0000	1.0000	1.0000	1.0000	1.0000	1.0000	1.0000	1.0000	1.0000	1.0000

TABLE A.2 (Continued)

n	k	p = .05	.10	.15	.20	.25	.30	.35	.40	.45
15	0	.4633	.2059	.0874	.0352	.0134	.0047	.0016	.0005	.0001
	1	.8290	.5490	.3186	.1671	.0802	.0353	.0142	.0052	.0017
	2	.9638	.8159	.6042	.3980	.2361	.1268	.0617	.0271	.0107
	3	.9945	.9444	.8227	.6482	.4613	.2969	.1727	.0905	.1424
	4	.9994	.9873	.9383	.8358	.6865	.5155	.3519	.2173	.1204
	5	.9999	.9978	.9832	.9389	.8516	.7216	.5643	.4032	.2608
	6	1.0000	.9997	.9964	.9819	.9434	.8689	.7548	.6098	.4522
	7	1.0000	1.0000	.9994	.9958	.9827	.9500	.8868	.7869	.6535
	8	1.0000	1.0000	.9999	.9992	.9958	.9848	.9578	.9050	.8121
	9	1.0000	1.0000	1.0000	.9999	.9992	.9963	.9876	.9662	.9231
	10	1.0000	1.0000	1.0000	1.0000	.9999	.9993	.9972	.9907	.9745
	11	1.0000	1.0000	1.0000	1.0000	1.0000	.9999	.9995	.9981	.9937
	12	1.0000	1.0000	1.0000	1.0000	1.0000	1.0000	.9999	.9997	.9989
	13	1.0000	1.0000	1.0000	1.0000	1.0000	1.0000	1.0000	1.0000	.9999
	14	1.0000	1.0000	1.0000	1.0000	1.0000	1.0000	1.0000	1.0000	1.0000
	15	1.0000	1.0000	1.0000	1.0000	1.0000	1.0000	1.0000	1.0000	1.0000
16	0	.4401	.1853	.0743	.0281	.0100	.0033	.0010	.0003	.0001
	1	.8108	.5147	.2839	.1407	.0635	.0261	.0098	.0088	.0010
	2	.9571	.7892	.5614	.3518	.1971	.0904	.0451	.0183	.0066
	3	.9930	.9316	.7899	.5981	.4050	.2459	.1339	.0651	.0281
	4	.9991	.9830	.9209	.7982	.6302	.4499	.2892	.1666	.0853
	5	.9999	.9967	.9765	.9183	.8103	.6598	.4900	.3288	.1970
	6	1.0000	.9995	.9944	.9733	.9204	.8247	.6881	.5272	.3660
	7	1.0000	.9999	.9989	.9930'	.9729	.9256	.8406	.7161	.5629
	8	1.0000	1.0000	.9998	.9985	.9925	.9743	.9329	.8577	.7441
	9	1.0000	1.0000	1.0000	.9998	.9984	.9938	.9809	.9514	.8759
	10	1.0000	1.0000	1.0000	1.0000	.9997	.9984	.9938	.9809	.9514
	11	1.0000	1.0000	1.0000	1.0000	1.0000	.9997	.9987	.9951	.9851
	12	1.0000	1.0000	1.0000	1.0000	1.0000	1.0000	.9998	.9991	.9965
	13	1.0000	1.0000	1.0000	1.0000	1.0000	1.0000	1.0000	.9999	.9965
	14	1.0000	1.0000	1.0000	1.0000	1.0000	1.0000	1.0000	1.0000	.9999
	15	1.0000	1.0000	1.0000	1.0000	1.0000	1.0000	1.0000	1.0000	1.0000
	16	1.0000	1.0000	1.0000	1.0000	1.0000	1.0000	1.0000	1.0000	1.0000

$p = .50$	.55	.60	.65	.70	.75	.80	.85	.90	.95
.0000	.0000	.0000	.0000	.0000	.0000	.0000	.0000	.0000	.0000
.0005	.0001	.0000	.0000	.0000	.0000	.0000	.0000	.0000	.0000
.0037	.0011	.0003	.0001	.0000	.0000	.0000	.0000	.0000	.0000
.0176	.0063	.0019	.0005	.0001	.0000	.0000	.0000	.0000	.0000
.0592	.0255	.0093	.0028	.0007	.0001	.0000	.0000	.0000	.0000
.1509	.1769	.0228	.0124	.0037	.0008	.0001	.0000	.0000	.0000
.3036	.1818	.0950	.0422	.0152	.0042	.0008	.0001	.0000	.0000
.5000	.3465	.2131	.1132	.0500	.0173	.0042	.0006	.0000	.0000
.6964	.5478	.3902	.2452	.1311	.0566	.0181	.0036	.0003	.0000
.8491	.7392	.5968	.4357	.2784	.1484	.0611	.0168	.0022	.0001
.9408	.8796	.7827	.6481	.4845	.3135	.1642	.0617	.0127	.0006
.9824	.9576	.9095	.8273	.7031	.5387	.3518	.1773	.0556	.0055
.9963	.9893	.9729	.9383	.8732	.7639	.6020	.3958	.1841	.0362
.9995	.9983	.9948	.9858	.9647	.9198	.8329	.6814	.4510	.1710
1.0000	.9999	.9995	.9984	.9953	.9866	.9648	.9126	.7941	.5367
1.0000	1.0000	1.0000	1.0000	1.0000	1.0000	1.0000	1.0000	1.0000	1.0000
.0000	.0000	.0000	.0000	.0000	.0000	.0000	.0000	.0000	.0000
.0003	.0001	.0000	.0000	.0000	.0000	.0000	.0000	.0000	.0000
.0021	.0006	.0001	.0000	.0000	.0000	.0000	.0000	.0000	.0000
.0106	.0035	.0009	.0002	.0000	.0000	.0000	.0000	.0000	.0000
.0384	.0149	.0049	.0013	.0003	.0000	.0000	.0000	.0000	.0000
.1051	.0486	.0191	.0062	.0016	.0003	.0000	.0000	.0000	.0000
.2272	.1241	.0583	.0229	.0071	.0016	.0002	.0000	.0000	.0000
.4018	.2559	.1423	.0671	.0257	.0075	.0015	.0002	.0000	.0000
.5982	.4371	.2839	.1594	.0744	.0271	.0070	.0011	.0001	.0000
.7728	.6340	.4728	.3119	.1753	.0796	.0267	.0056	.0005	.0000
.8949	.8024	.6712	.5100	.3402	.1897	.0817	.0235	.0033	.0001
.9616	.9147	.8334	.7108	.5501	.3698	.2018	.0791	.0170	.0009
.9894	9719	.9349	.8661	.7541	.5950	.4019	.2101	.0684	.0070
.9979	.9934	.9817	.9549	.9006	.8729	.6482	.4386	.2108	.0429
.9997	.9990	.9967	.9902	.9739	.9365	.8593	.7176	.4853	.1892
1.0000	.9999	.9997	.9990	.9967	.9900	.9719	.9257	.8147	.5599
1.0000	1.0000	1.0000	1.0000	1.0000	1.0000	1.0000	1.0000	1.0000	1.0000

TABLE A.2 (Continued)

n	k	p = .05	.10	.15	.20	.25	.30	.35	.40	.45
17	0	.4181	.1668	.0631	.0225	.0075	.0023	.0007	.0002	.0000
	1	.7922	.4818	.2525	.1182	.0501	.0193	.0067	.0021	.0006
	2	.9497	.7618	.5198	.3096	.1637	.0774	.0327	.0123	.0041
	3	.9912	.9174	.7556	.5489	.3530	.2019	.1028	.0464	.0184
	4	.9988	.9779	.9013	.7582	.5739	.3887	.2348	.1260	.0596
	5	.9999	.9953	.9681	.8943	.7653	.5968	.4197	.2639	.1471
	6	1.0000	.9992	.9917	.9623	.8929	.7752	.6188	.4478	.2902
	7	1.0000	.9999	.9983	.9891	.9598	.8954	.7872	.6405	.4743
	8	1.0000	1.0000	.9997	.9974	.9876	.9597	.9006	.8011	.6626
	9	1.0000	1.0000	1.0000	.9995	.9969	.9873	.9611	.9081	.8166
	10	1.0000	1.0000	1.0000	.9999	.9994	.9968	.9880	.9652	.9174
	11	1.0000	1.0000	1.0000	1.0000	.9999	.9993	.9970	.9894	.9699
	12	1.0000	1.0000	1.0000	1.0000	1.0000	.9999	.9994	.9975	.9914
	13	1.0000	1.0000	1.0000	1.0000	1.0000	1.0000	.9999	.9995	.9981
	14	1.0000	1.0000	1.0000	1.0000	1.0000	1.0000	1.0000	.9999	.9997
	15	1.0000	1.0000	1.0000	1.0000	1.0000	1.0000	1.0000	1.0000	1.0000
	16	1.0000	1.0000	1.0000	1.0000	1.0000	1.0000	1.0000	1.0000	1.0000
	17	1.0000	1.0000	1.0000	1.0000	1.0000	1.0000	1.0000	1.0000	1.0000
18	0	3972	.1501	.0536	.0180	.0056	.0016	.0004	.0001	.0000
	1	.7735	.4503	.2241	.0991	.0395	.0142	.0046	.0013	.0003
	2	.9419	.7338	.4797	.2713	.1353	.0600	.0236	.0082	.0025
	3	.9891	.9018	.7202	.5010	.3057	.1646	.0783	.0328	.0120
	4	.9985	.9718	.8794	.7164	.5187	.3327	.1886	.0942	.0411
	5	.9998	.9936	.9581	.8671	.7175	.5344	.3550	.2088	.1077
	6	1.0000	.9988	.9973	.9837	.9431	.8593	.7283	.5634	.3915
	7	1.0000	.9998	.9973	.9837	.9431	.8593	.7283	.5634	.3915
	8	1.0000	1.0000	.9995	.9957	.9807	.9404	.8609	.7368	.5778
	9	1.0000	1.0000	.9999	.9991	.9946	.9790	.9403	.8653	.7473
	10	1.0000	1.0000	1.0000	.9998	.9988	.9939	.9788	.9424	.8720
	11	1.0000	1.0000	1.0000	1.0000	.9998	.9986	.9938	.9797	.9463
	12	1.0000	1.0000	1.0000	1.0000	1.0000	.9997	.9986	.9942	.9817
	13	1.0000	1.0000	1.0000	1.0000	1.0000	1.0000	.9997	.9987	.9951
	14	1.0000	1.0000	1.0000	1.0000	1.0000	1.0000	1.0000	.9998	.9990
	15	1.0000	1.0000	1.0000	1.0000	1.0000	1.0000	1.0000	1.0000	.9999
	16	1.0000	1.0000	1.0000	1.0000	1.0000	1.0000	1.0000	1.0000	1.0000
	17	1.0000	1.0000	1.0000	1.0000	1.0000	1.0000	1.0000	1.0000	1.0000
	18	1.0000	1.0000	1.0000	1.0000	1.0000	1.0000	1.0000	1.0000	1.0000

p = .50	.55	.60	.65	.70	.75	.80	.85	.90	.95
.0000	.0000	.0000	.0000	.0000	.0000	.0000	.0000	.0000	.0000
.0001	.0000	.0000	.0000	.0000	.0000	.0000	.0000	.0000	.0000
.0012	.0003	.0001	.0000	.0000	.0000	.0000	.0000	.0000	.0000
.0064	.0019	.0005	.0001	.0000	.0000	.0000	.0000	.0000	.0000
.0245	.0086	.0025	.0006	.0001	.0000	.0000	.0000	.0000	.0000
.0717	.0301	.0106	.0030	.0007	.0001	.0000	.0000	.0000	.0000
.1662	.0826	.0348	.0120	.0032	.0006	.0001	.0000	.0000	.0000
.3145	.1834	.0919	.0383	.0127	.0031	.0005	.0000	.0000	.0000
.5000	.3374	.1989	.0994	.0403	.0124	.0026	.0003	.0000	.0000
.6855	.5257	.3595	.2128	.1046	.0402	.0109	.0017	.0001	.0000
.8338	.7098	.5522	.3812	.2248	.1071	.0377	.0083	.0008	.0000
.9283	.8529	.7361	.5803	.4032	.2347	.1057	.0319	.0047	.0001
.9755	.9404	.8740	.7652	.6113	.4261	.2418	.0987	.0221	.0012
.9936	.9816	.9536	.8972	.7981	.6470	.4511	.2444	.0826	.0088
.9988	.9959	.9877	.9673	.9226	.8363	.6904	.4802	.2382	.0503
.9999	.9994	.9979	.9933	.9807	.9499	.8818	.7475	.5182	.2078
1.0000	1.0000	.9998	.0003	.0077	.9925	.9775	.9369	.8332	.5819
1.0000	1.0000	1.0000	1.0000	1.0000	1.0000	1.0000	1.0000	1.0000	1.0000
.0000	.0000	.0000	.0000	.0000	.0000	.0000	.0000	.0000	.0000
.0001	.0000	.0000	.0000	.0000	.0000	.0000	.0000	.0000	.0000
.0007	.0001	.0000	.0000	.0000	.0000	.0000	.0000	.0000	.0000
.0038	.0010	.0002	.0000	.0000	.0000	.0000	.0000	.0000	.0000
.0154	.0049	.0013	.0003	.0000	.0000	.0000	.0000	.0000	.0000
.0481	.0183	.0058	.0014	.0003	.0000	.0000	.0000	.0000	.0000
.1189	.0537	.0203	.0062	.0014	.0002	.0000	.0000	.0000	.0000
.2403	.1280	.0576	.0212	.0061	.0012	.0002	.0000	.0000	.0000
.4073	.2527	.1347	.0597	.0210	.0054	.0009	.0001	.0000	.0000
.5927	.4222	.2632	.2717	.1407	.0569	.0163	.0027	.0002	.0000
.7597	.6085	.4366	.2717	.1407	.0569	.0163	.0027	.0002	.0000
.8811	.7742	.6457	.4509	.2783	.1390	.0513	.0118	.0012	.0000
.9519	.8923	.7912	.6450	.4656	.2825	.1329	.0419	.0064	.0002
.9846	.9589	.9058	.8114	.6673	.4813	.2836	.1206	.0282	.0015
.9962	.9880	.9672	.9217	.8354	.6943	.4990	.2798	.0982	.0109
.9993	.9975	.9918	.9764	.9400	.8647	.7287	.5203	.2662	.0581
.9999	.9997	.9987	.9954	.9858	.9605	.9009	.7759	.5497	.2265
1.0000	1.0000	.9999	.9996	.9984	.9944	.9820	.9464	.8499	.6028
1.0000	1.0000	1.0000	1.0000	1.0000	1.0000	1.0000	1.0000	1.0000	1.0000

TABLE A.2 (Continued)

n	k	p = .05	.10	.15	.20	.25	.30	.35	.40	.45
19	0	.3774	.1351	.0456	.0144	.0042	.0011	.0003	.0001	.0000
	1	.7547	.4203	.1985	.0829	.0310	.0008	.0002	.0008	.0002
	2	.9335	.7054	.4413	.2369	.1113	.0462	.0170	.0055	.0015
	3	.9869	.8850	.6841	.4551	.2631	.1332	.0591	.0230	.0077
	4	.9869	.8850	.6841	.4551	.2631	.1332	.0591	.0230	.0077
	5	.9998	.9914	.9463	.8369	.6678	.4739	.2968	.1629	.0777
	6	1.0000	.9983	.9837	.9324	.8251	.6655	.4912	.3081	.1727
	7	1.0000	.9997	.9959	.9767	.9225	.8180	.6656	.4878	.3169
	8	1.0000	1.0000	.9992	.9933	.9713	.9161	.8145	.6675	.4940
	9	1.0000	1.0000	.9999	.9984	.9911	.9674	.9125	.8139	.6710
	10	1.0000	1.0000	1.0000	.9997	.9977	.9895	.9653	.9115	.8159
	11	1.0000	1.0000	1.0000	1.0000	.9995	.9972	.9886	.9648	.9129
	12	1.0000	1.0000	1.0000	1.0000	.9999	.9994	.9969	.9884	.9658
	13	1.0000	1.0000	1.0000	1.0000	1.0000	.9999	.9993	.9969	.9891
	14	1.0000	1.0000	1.0000	1.0000	1.0000	1.0000	.9999	.9994	.9972
	15	1.0000	1.0000	1.0000	1.0000	1.0000	1.0000	1.0000	.9999	.9995
	16	1.0000	1.0000	1.0000	1.0000	1.0000	1.0000	1.0000	1.0000	.9999
	17	1.0000	1.0000	1.0000	1.0000	1.0000	1.0000	1.0000	1.0000	1.0000
	18	1.0000	1.0000	1.0000	1.0000	1.0000	1.0000	1.0000	1.0000	1.0000
	19	1.0000	1.0000	1.0000	1.0000	1.0000	1.0000	1.0000	1.0000	1.0000
20	0	.3585	.1261	.0388	.0115	.0032	.0008	.0000	.0000	.0000
	1	.7358	.3917	.1756	.0692	.0243	.0076	.0021	.0005	.0001
	2	.9245	.6769	.4049	.2061	.0913	.0355	.0121	.0036	.0009
	3	.9841	.8670	.6477	.4114	.2252	.1071	.0444	.0160	.0049
	4	.9974	.9568	.8298	.6296	.4148	.2375	.1182	.0510	.0189
	5	.9997	.9887	.9327	.8042	.6172	.4164	.2454	.1256	.0553
	6	1.0000	.9976	.9781	.9133	.7858	.6080	.4166	.2500	.1299
	7	1.0000	.9996	.9941	.9679	.8982	.7723	.6010	.4159	.2520
	8	1.0000	.9999	.9987	.9900	.9591	.8867	.7624	.5956	.4143
	9	1.0000	1.0000	.9998	.9974	.9861	.9520	.8782	.7553	.5914
	10	1.0000	1.0000	1.0000	.9994	.9961	.9829	.9468	.8725	.7507
	11	1.0000	1.0000	1.0000	.9999	.9991	.9949	.9804	.9435	.8692
	12	1.0000	1.0000	1.0000	1.0000	.9998	.9987	.9940	.9790	.9420
	13	1.0000	1.0000	1.0000	1.0000	1.0000	.9997	.9985	.9935	.9786
	14	1.0000	1.0000	1.0000	1.0000	1.0000	1.0000	.9997	.9984	.9936
	15	1.0000	1.0000	1.0000	1.0000	1.0000	1.0000	1.0000	.9997	.9985
	16	1.0000	1.0000	1.0000	1.0000	1.0000	1.0000	1.0000	1.0000	.9997
	17	1.0000	1.0000	1.0000	1.0000	1.0000	1.0000	1.0000	1.0000	1.0000
	18	1.0000	1.0000	1.0000	1.0000	1.0000	1.0000	1.0000	1.0000	1.0000
	19	1.0000	1.0000	1.0000	1.0000	1.0000	1.0000	1.0000	1.0000	1.0000
	20	1.0000	1.0000	1.0000	1.0000	1.0000	1.0000	1.0000	1.0000	1.0000

$p = .50$	.55	.60	.65	.70	.75	.80	.85	.90	.95
.0000	.0000	.0000	.0000	.0000	.0000	.0000	.0000	.0000	.0000
.0000	.0000	.0000	.0000	.0000	.0000	.0000	.0000	.0000	.0000
.0004	.0001	.0000	.0000	.0000	.0000	.0000	.0000	.0000	.0000
.0022	.0005	.0001	.0000	.0000	.0000	.0000	.0000	.0000	.0000
.0096	.0028	.0006	.0001	.0000	.0000	.0000	.0000	.0000	.0000
.0318	.0109	.0031	.0007	.0001	.0000	.0000	.0000	.0000	.0000
.0835	.0342	.0116	.0031	.0006	.0001	.0000	.0000	.0000	.0000
.1796	.0871	.0352	.0114	.0028	.0005	.0000	.0000	.0000	.0000
.3238	.1841	.0885	.0347	.0105	.0023	.0003	.0000	.0000	.0000
.5000	.3290	.1861	.0875	.0326	.0287	.0067	.0008	.0000	.0000
.6762	.5060	.3325	.1855	.0839	.0287	.0067	.0008	.0000	.0000
.8204	.6831	.5122	.3344	.1820	.0775	.0233	.0041	.0008	.0000
.9165	.8273	.6919	.5188	.3345	.1749	.0676	.0163	.0017	.0000
.9682	.9223	.8371	.7032	.5261	.3322	.1631	.0537	.0086	.0002
.9904	.9720	.9304	.8500	.7178	.5346	.3267	.1444	.0352	.0020
.9978	.9923	.9770	.9409	.8668	.7369	.5449	.3159	.1150	.0132
.9996	.9985	.9945	.9830	.9538	.8887	.7631	.5587	.2946	.0665
1.0000	.9998	.9992	.9969	.9896	.9690	.9171	.8015	.5797	.2453
1.0000	1.0000	.9999	.9997	.9989	.9958	.9856	.9544	.8649	.6226
1.0000	1.0000	1.0000	1.0000	1.0000	1.0000	1.0000	1.0000	1.0000	1.0000
.0000	.0000	.0000	.0000	.0000	.0000	.0000	.0000	.0000	.0000
.0000	.0000	.0000	.0000	.0000	.0000	.0000	.0000	.0000	.0000
.0002	.0000	.0000	.0000	.0000	.0000	.0000	.0000	.0000	.0000
.0013	.0003	.0000	.0000	.0000	.0000	.0000	.0000	.0000	.0000
.0059	.0015	.0003	.0000	.0000	.0000	.0000	.0000	.0000	.0000
.0207	.0064	.0016	.0003	.0000	.0000	.0000	.0000	.0000	.0000
.0577	.0214	.0065	.0015	.0003	.0000	.0000	.0000	.0000	.0000
.1316	.0580	.0210	.0060	.0013	.0002	.0000	.0000	.0000	.0000
.2517	.1308	.0565	.0196	.0051	.0009	.0001	.0000	.0000	.0000
.4119	.2493	.1275	.0532	.0171	.0039	.0006	.0000	.0000	.0000
.5881	.4086	.2447	.1218	.0480	.0139	.0026	.0002	.0000	.0000
.7483	.5857	.4044	.2376	.1133	.0409	.0100	.0013	.0001	.0000
.8684	.7480	.5841	.3990	.2277	.1018	.0321	.0059	.0004	.0000
.9423	.8701	.7500	.5834	.3920	.2142	.0867	.0219	.0024	.0000
.9793	.9447	.8744	.7546	.5836	.3828	.1958	.0673	.0113	.0003
.9941	.9811	.9490	.8818	.7625	.5852	.3704	.1702	.0432	.0026
.9987	.9951	.9840	.9556	.8929	.7748	.5886	.3523	.1330	.0159
.9998	.9991	.9964	.9879	.9645	.9087	.7939	.5951	.3231	.0755
1.0000	.9999	.9995	.9979	.9924	.9757	.9308	.8244	.6083	.2642
1.0000	1.0000	1.0000	.9998	.9992	.9968	.9885	.9612	.8784	.6415
1.0000	1.0000	1.0000	1.0000	1.0000	1.0000	1.0000	1.0000	1.0000	1.0000

SOURCE: M. Orkin and R. Drogin, "Vital Statistics," pp. 338–349, McGraw-Hill, New York, 1975. I would like to thank W. J. Conover for calling my attention to a small number of minor errors in an earlier version of this table, which had previously appeared in his book "Practical Nonparametric Statistics," Wiley, New York, 1971.

Table A.3 gives the proportion to the left of z of a standard normal set of data, for various values of z.

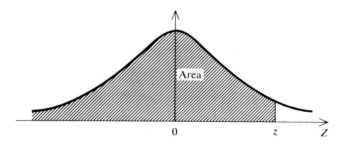

To obtain proportions use the fact that

Proportion to be left of $z = P(Z \leq z)$

where Z is a random variable with a standard normal distribution.

To find standerd deviation: - calculate deviations x-μ
                        = Square each (x each by itself)
                        - Σ squared deviations

- ÷ Σ by n
- Take square root of resulting #.

- Chebysher's Theorem : $\mu - k\sigma$ or $\mu + k\sigma$ 1/k²
- z scores (standard scores) ⟹ $z = \dfrac{x - \mu}{\sigma}$

—⟶ ⟍   —⟶ ⟍   —⟶ ⟍   —⟶ ⟍   —⟶ ⟍   —⟶ ⟍   —⟶ ⟍

Chpt. 2
   - Binomial Distrbution - p; 1-p 3 outcomes   n = # times experiment repeated
    formula ⟹ $P(H=K) = \dfrac{n!}{K!(n-K)!} \, p^{K}(1-p)^{n-K}$

    $\mu = np$
    $\sigma = \sqrt{np(1-p)}$
   - Normal Distribution - characteristics $\mu = 0$ & $\sigma = 1$

TABLE A.3
Standard normal distribution

z	Proportion	z	Proportion	z	Proportion
−4	.00003	−2.74	.0031	−2.29	.0110
−3.9	.00005	−2.73	.0032	−2.28	.0113
−3.8	.0001	−2.72	.0033	−2.27	.0116
−3.7	.0001	−2.71	.0034	−2.26	.0119
−3.6	.0002	−2.70	.0035	−2.25	.0122
−3.5	.0002	−2.69	.0036	−2.24	.0125
−3.4	.0003	−2.68	.0037	−2.23	.0129
−3.3	.0005	−2.67	.0038	−2.22	.0132
−3.2	.0007	−2.66	.0039	−2.21	.0136
−3.1	.0010	−2.65	.0040	−2.20	.0139
−3.09	.0010	−2.64	.0041	−2.19	.0143
−3.08	.0010	−2.63	.0043	−2.18	.0146
−3.07	.0011	−2.62	.0044	−2.17	.0150
−3.06	.0011	−2.61	.0045	−2.16	.0154
−3.05	.0011	−2.60	.0047	−2.15	.0158
−3.04	.0012	−2.59	.0048	−2.14	.0162
−3.03	.0012	−2.58	.0049	−2.13	.0166
−3.02	.0013	−2.57	.0051	−2.12	.0170
−3.01	.0013	−2.56	.0052	−2.11	.0174
−3.00	.0013	−2.55	.0054	−2.10	.0179
−2.99	.0014	−2.54	.0055	−2.09	.0183
−2.98	.0014	−2.53	.0057	−2.08	.0188
−2.97	.0015	−2.52	.0059	−2.07	.0182
−2.96	.0015	−2.51	.0060	−2.06	.0197
−2.95	.0016	−2.50	.0062	−2.05	.0202
−2.94	.0016	−2.49	.0064	−2.04	.0207
−2.93	.0017	−2.48	.0066	−2.03	.0212
−2.92	.0017	−2.47	.0068	−2.02	.0217
−2.91	.0018	−2.46	.0069	−2.01	.0222
−2.90	.0019	−2.45	.0071	−2.00	.0228
−2.89	.0019	−2.44	.0073	−1.99	.0233
−2.88	.0020	−2.43	.0075	−1.98	.0239
−2.87	.0021	−2.42	.0078	−1.97	.0244
−2.86	.0021	−2.41	.0080	−1.96	.0250
−2.85	.0022	−2.40	.0082	−1.95	.0256
−2.84	.0023	−2.39	.0084	−1.94	0262
−2.83	.0023	−2.38	.0087	−1.93	.0268
−2.82	.0024	−2.37	.0089	1.92	.0274
−2.81	.0025	−2.36	.0091	−1.91	.0281
−2.80	.0026	−2.35	.0094	−1.90	.0287
−2.79	.0026	−2.34	.0096	−1.89	.0294
−2.78	.0027	−2.33	.0099	−1.88	.030
−2.77	.0028	−2.32	.0102	−1.87	.0307
−2.76	.0029	−2.31	.0104	−1.86	.0314
−2.75	.0030	−2.30	.0107	−1.85	.0322

Chpt·1 — Frequecy Distribution — divide data into intervals & count data pts.
sum # data pts.

Histogram (bars) ⇒ ex 19.5-22.5
Frequency polygon ⇒ determined by midpt. ⇒ ex. 20-22 mdpt. 21

- Averages: median located in middle
mode is # that appears the most.
μ = mean is sum of #'s ÷ how many there are, ⇒ $\mu = \frac{\Sigma x}{n}$
deviation from mean ⇒ $x - \mu$ = standard di.
mean absolute deviation = $\frac{\text{sum of stand. Der}}{\text{number of data pts}}$ ⇒ $\frac{\Sigma |x - \mu|}{n}$

standard deviation =

$$\sigma = \sqrt{\frac{\Sigma (x - \mu)^2}{n}}$$

z	Proportion	z	Proportion	z	Proportion
−1.84	.0329	−1.39	.0823	− .94	.1736
−1.83	.0336	−1.38	.0838	− .93	.1762
−1.82	.0344	−1.37	.0853	− .92	.1788
−1.81	.0352	−1.36	.0869	− .91	.1814
−1.80	.0359	−1.35	.0885	− .90	.1841
−1.79	.0367	−1.34	.0901	− .89	.1867
−1.78	.0375	−1.33	.0918	− .88	.1894
−1.77	.0384	−1.32	.0934	− .87	.1922
−1.76	.0392	−1.31	.0951	− .86	.1949
−1.75	.0401	−1.30	.0968	− .85	.1977
−1.74	.0409	−1.29	.0985	− .84	.2005
−1.73	.0418	−1.28	.1003	− .83	.2033
−1.72	.0427	−1.27	.1020	− .82	.2061
−1.71	.0436	−1.26	.1038	− .81	.2090
−1.70	.0446	−1.25	.1056	− .80	.2119
−1.69	.0455	−1.24	.1075	− .79	.2148
−1.68	.0465	−1.23	.1093	− .78	.2177
−1.67	.0475	−1.22	.1112	− .77	.2296
−1.66	.0485	−1.21	.1131	− .76	.2236
−1.65	.0495	−1.20	.1151	− .75	.2266
−1.64	.0505	−1.19	.1170	− .74	.2296
−1.63	.0516	−1.18	.1190	− .73	.2327
−1.62	.0526	−1.17	.1210	− .72	.2358
−1.61	.0537	−1.16	.1230	− .71	.2389
−1.60	.0548	−1.15	.1251	− .70	.2420
−1.59	.0559	−1.14	.1271	− .69	.2451
−1.58	.0571	−1.13	.1292	− .68	.2483
−1.57	.0582	−1.12	.1314	− .67	.2514
−1.56	.0594	−1.11	.1335	− .66	.2546
−1.55	.0606	−1.10	.1357	− .65	.2578
−1.54	.0618	−1.09	.1379	− .64	.2611
−1.53	.0630	−1.08	.1401	− .63	.2643
−1.52	.0643	−1.07	.1423	− .62	.2676
−1.51	.0655	−1.06	.1446	− .61	.2709
−1.50	.0668	−1.05	.1469	− .60	.2743
−1.49	.0681	−1.04	.1492	− .59	.2776
−1.48	.0694	−1.03	.1515	− .58	.2810
−1.47	.0708	−1.02	.1539	− .57	.2843
−1.46	.0722	−1.01	.1562	− .56	.2877
−1.45	.0735	−1.00	.1587	− .55	.2912
−1.44	.0749	− .99	.1611	− .54	.2946
−1.43	.0764	− .98	.1635	− .53	.2981
−1.42	.0778	− .97	.1660	− .52	.3015
−1.41	.0793	− .96	.1685	− .51	.3050
−1.40	.0808	− .95	.1711	− .50	.3085

TABLE A.3 (Continued)

z	Proportion	z	Proportion	z	Proportion
− .49	.3121	− .04	.4840	.41	.6591
− .48	.3156	− .03	.4880	.42	.6628
− .47	.3192	− .02	.4920	.43	.6664
− .46	.3228	− .01	.4960	.44	.6700
− .45	.3264	.00	.5000	.45	.6736
− .44	.3300	.01	.5040	.46	.6772
− .43	.3336	.02	.5080	.47	.6808
− .42	.3372	.03	.5120	.48	.6844
− .41	.3409	.04	.5160	.49	.6849
− .40	.3446	.05	.5199	.50	.6915
− .39	.3483	.06	.5239	.51	.6950
− .38	.3520	.07	.5279	.52	.6985
− .37	.3557	.08	.5319	.53	.7019
− .36	.3594	.09	.5359	.54	.7054
− .35	.3632	.10	.5398	.55	.7088
− .34	.3669	.11	.5438	.56	.7123
− .33	.3707	.12	.5478	.57	.7157
− .32	.3745	.13	.5517	.58	.7190
− .31	.3783	.14	.5557	.59	.7224
− .30	.3821	.15	.5596	.60	.7257
− .29	.3859	.16	.5636	.61	.7291
− .28	.3897	.17	.5675	.62	.7324
− .27	.3936	.18	.5714	.63	.7357
− .26	.3974	.19	.5753	.64	.7389
− .25	.4013	.20	.5793	.65	.7422
− .24	.4052	.21	.5832	.66	.7454
− .23	.4090	.22	.5871	.67	.7486
− .22	.4129	.23	.5910	.68	.7517
− .21	.4168	.24	.5948	.69	.7549
− .20	.4207	.25	.5987	.70	.7580
− .19	.4247	.26	.6026	.71	.7611
− .18	.4286	.27	.6064	.72	.7642
− .17	.4325	.28	.6103	.73	.7673
− .16	.4364	.29	.6141	.74	.7704
− .15	.4404	.30	.6179	.75	.7734
− .14	.4443	.31	.6217	.76	.7764
− .13	.4483	.32	.6255	.77	.7794
− .12	.4522	.33	.6293	.78	.7823
− .11	.4562	.34	.6331	.79	.7852
− .10	.4602	.35	.6368	.80	.7881
− .09	.4641	.36	.6406	.81	.7910
− .08	.4681	.37	.6443	.82	.7939
− .07	.4721	.38	.6480	.83	.7967
− .06	.4761	.39	.6517	.84	.7995
− .05	.4801	.40	.6554	.85	.8023

z	Proportion	z	Proportion	z	Proportion
.86	.8051	1.31	.9049	1.76	.9608
.87	.8078	1.32	.9066	1.77	.9616
.88	.8106	1.33	.9082	1.78	.9625
.89	.8133	1.34	.9099	1.79	.9633
.90	.8159	1.35	.9115	1.80	.9641
.91	.8186	1.36	.9131	1.81	.9649
.92	.8212	1.37	.9147	1.82	.9656
.93	.8238	1.38	.9162	1.83	.9664
.94	.8264	1.39	.9177	1.84	.9671
.95	.8289	1.40	.9192	1.85	.9678
.96	.8315	1.41	.9207	1.86	.9686
.97	.8340	1.42	.9222	1.87	.9693
.98	.8365	1.43	.9236	1.88	.9699
.99	.8389	1.44	.9251	1.89	.9706
1.00	.8413	1.45	.9265	1.90	.9713
1.01	.8438	1.46	.9278	1.91	.9719
1.02	.8461	1.47	.9292	1.92	.9726
1.03	.8485	1.48	.9306	1.93	.9732
1.04	.8508	1.49	.9319	1.94	.9738
1.05	.8531	1.50	.9332	1.95	.9744
1.06	.8554	1.51	.9345	1.96	.9750
1.07	.8577	1.52	.9357	1.97	.9756
1.08	.8599	1.53	.9370	1.98	.9761
1.09	.8621	1.54	.9382	1.99	.9767
1.10	.8643	1.55	.9394	2.00	.9772
1.11	.8665	1.56	.9406	2.01	.9778
1.12	.8686	1.57	.9418	2.02	.9783
1.13	.8708	1.58	.9429	2.03	.9788
1.14	.8729	1.59	.9441	2.04	.9793
1.15	.8749	1.60	.9452	2.05	.9798
1.16	.8770	1.61	.9463	2.06	.9803
1.17	.8790	1.62	.9474	2.07	.9808
1.18	.8810	1.63	.9484	2.08	.9812
1.19	.8830	1.64	.9495	2.09	.9817
1.20	.8849	1.65	.9505	2.10	.9821
1.21	.8869	1.66	.9515	2.11	.9826
1.22	.8888	1.67	.9525	2.12	.9830
1.23	.8907	1.68	.9535	2.13	.9834
1.24	.8925	1.69	.9545	2.14	.9838
1.25	.8944	1.70	.9554	2.15	.9842
1.26	.8962	1.71	.9564	2.16	.9846
1.27	.8980	1.72	.9573	2.17	.9850
1.28	.8997	1.73	.9582	2.18	.9854
1.29	.9015	1.74	.9591	2.19	.9857
1.30	.9032	1.75	.9599	2.20	.9861

TABLE A.3 **(Continued)**

z	Proportion	z	Proportion	z	Proportion
2.21	.9864	2.66	.9961	3.2	.9993
2.22	.9868	2.67	.9962	3.3	.9995
2.23	.9871	2.68	.9963	3.4	.9997
2.24	.9875	2.69	.9964	3.5	.9998
2.25	.9878	2.70	.9965	3.6	.9998
				3.7	.9999
2.26	.9881	2.71	.9966	3.8	.9999
2.27	.9884	2.72	.9967		
2.28	.9887	2.73	.9968		
2.29	.9890	2.74	.9969	3.9	.99995
2.30	.9893	2.75	.9970	4.0	.99997
2.31	.9896	2.76	.9971		
2.32	.9898	2.77	.9972		
2.33	.9901	2.78	.9973		
2.34	.9904	2.79	.9974		
2.35	.9906	2.80	.9974		
2.36	.9909	2.81	.9975		
2.37	.9911	2.82	.9976		
2.38	.9913	2.83	.9977		
2.39	.9916	2.84	.9977		
2.40	.9918	2.85	.9978		
2.41	.9920	2.86	.9979		
2.42	.9922	2.87	.9979		
2.43	.9925	2.88	.9980		
2.44	.9927	2.89	.9981		
2.45	.9929	2.90	.9981		
2.46	.9931	2.91	.9982		
2.47	.9932	2.92	.9982		
2.48	.9934	2.93	.9983		
2.49	.9936	2.94	.9984		
2.50	.9938	2.95	.9984		
2.51	.9940	2.96	.9985		
2.52	.9941	2.97	.9985		
2.53	.9943	2.98	.9986		
2.54	.9945	2.99	.9986		
2.55	.9946	3.00	.9987		
2.56	.9948	3.01	.9987		
2.57	.9949	3.02	.9987		
2.58	.9951	3.03	.9988		
2.59	.9952	3.04	.9988		
2.60	.9953	3.05	.9989		
2.61	.9955	3.06	.9989		
2.62	.9956	3.07	.9989		
2.63	.9957	3.08	.9990		
2.64	.9959	3.09	.9990		
2.65	.9960	3.10	.9990		

SOURCE: M. Orkin and R. Drogin, "Vital Statistics," pp. 350–355, McGraw-Hill, New York, 1975.

TABLE A.4
The _t_ distribution (values of t_α)*

df	$t_{.100}$	$t_{.050}$	$t_{.025}$	$t_{.010}$	$t_{.005}$	df
1	3.078	6.314	12.706	31.821	63.657	1
2	1.886	2.920	4.303	6.965	9.925	2
3	1.638	2.353	3.182	4.541	5.841	3
4	1.533	2.132	2.776	3.747	4.604	4
5	1.476	2.015	2.571	3.365	4.032	5
6	1.440	1.943	2.447	3.143	3.707	6
7	1.415	1.895	2.365	2.998	3.499	7
8	1.397	1.860	2.306	2.896	3.355	8
9	1.383	1.833	2.262	2.821	3.250	9
10	1.372	1.812	2.228	2.764	3.169	10
11	1.363	1.796	2.201	2.718	3.106	11
12	1.356	1.782	2.179	2.681	3.055	12
13	1.350	1.771	2.160	2.650	3.012	13
14	1.345	1.761	2.145	2.624	2.977	14
15	1.341	1.753	2.131	2.602	2.947	15
16	1.337	1.746	2.120	2.583	2.921	16
17	1.333	1.740	2.110	2.567	2.898	17
18	1.330	1.734	2.101	2.552	2.878	18
19	1.328	1.729	2.093	2.539	2.861	19
20	1.325	1.725	2.086	2.528	2.845	20
21	1.323	1.721	2.080	2.518	2.831	21
22	1.321	1.717	2.074	2.508	2.819	22
23	1.319	1.714	2.069	2.500	2.807	23
24	1.318	1.711	2.064	2.492	2.797	24
25	1.316	1.708	2.060	2.485	2.787	25
26	1.315	1.706	2.056	2.479	2.779	26
27	1.314	1.703	2.052	2.473	2.771	27
28	1.313	1.701	2.048	2.467	2.763	28
29	1.311	1.699	2.045	2.462	2.756	29
inf.	1.282	1.645	1.960	2.326	2.576	inf.

*This table is abridged from Table IV of R. A. Fisher, "Statistical Methods for Research Workers," copyright 1972 by Hafner Press, and is reproduced with permission of the Hafner Press.

TABLE A.5
The chi-square distribution (values of χ_α^2)*

df	$\chi_{.05}^2$	$\chi_{.025}^2$	$\chi_{.01}^2$	$\chi_{.005}^2$	df
1	3.841	5.024	6.635	7.879	1
2	5.991	7.378	9.210	10.597	2
3	7.815	9.348	11.345	12.838	3
4	9.488	11.143	13.277	14.860	4
5	11.071	12.833	15.086	16.750	5
6	12.592	14.449	16.812	18.548	6
7	14.067	16.013	18.475	20.278	7
8	15.507	17.535	20.090	21.955	8
9	16.919	19.023	21.666	23.589	9
10	18.307	20.483	23.209	25.188	10
11	19.675	21.920	24.725	26.757	11
12	21.026	23.337	26.217	28.300	12
13	22.362	24.736	27.688	29.820	13
14	23.685	26.119	29.141	31.319	14
15	24.996	27.488	30.578	32.801	15
16	26.296	28.845	32.000	34.267	16
17	27.587	30.191	33.409	35.719	17
18	28.869	31.526	34.805	37.157	18
19	30.144	32.852	36.191	38.582	19
20	31.410	34.170	37.566	39.997	20
21	32.671	35.479	38.932	41.401	21
22	33.924	36.781	40.289	42.796	22
23	35.173	38.076	41.638	44.181	23
24	36.415	39.364	42.980	45.559	24
25	37.653	40.647	44.314	46.928	25
26	38.885	41.923	45.642	48.290	26
27	40.113	43.195	46.963	49.645	27
28	41.337	44.461	48.278	50.993	28
29	42.557	45.722	49.588	52.336	29
30	43.773	46.979	50.892	53.672	30

*This table is abridged from table 8, page 137, of "Biometrika Tables for Statisticians," vol. 1, edited by E. S. Pearson and H. O. Hartley, published by Cambridge University Press, 1966, and is reproduced with permission of the "Biometrika" trustees.

TABLE A.6
The *F* distribution (values of $F_{.05}$)*

DEGREES OF FREEDOM FOR NUMERATOR

	1	2	3	4	5	6	7	8	9	10	12	15	20	24	30	40	60	120	∞
1	161	200	216	225	230	234	237	239	241	242	244	246	248	249	250	251	252	253	254
2	18.5	19.0	19.2	19.2	19.3	19.3	19.4	19.4	19.4	19.4	19.4	19.4	19.4	19.5	19.5	19.5	19.5	19.5	19.5
3	10.1	9.55	9.28	9.12	9.01	8.94	8.89	8.85	8.81	8.79	8.74	8.70	8.66	8.64	8.62	8.59	8.57	8.55	8.53
4	7.71	6.94	6.59	6.39	6.26	6.16	6.09	6.04	6.00	5.96	5.91	5.86	5.80	5.77	5.75	5.72	5.69	5.66	5.63
5	6.61	5.79	5.41	5.19	5.05	4.95	4.88	4.82	4.77	4.74	4.68	4.62	4.56	4.53	4.50	4.46	4.43	4.40	4.37
6	5.99	5.14	4.76	4.53	4.39	4.28	4.21	4.15	4.10	4.06	4.00	3.94	3.87	3.84	3.81	3.77	3.74	3.70	3.67
7	5.59	4.74	4.35	4.12	3.97	3.87	3.79	3.73	3.68	3.64	3.57	3.51	3.44	3.41	3.38	3.34	3.30	3.27	3.23
8	5.32	4.46	4.07	3.84	3.69	3.58	3.50	3.44	3.39	3.35	3.28	3.22	3.15	3.12	3.08	3.04	3.01	2.97	2.93
9	5.12	4.26	3.86	3.63	3.48	3.37	3.29	3.23	3.18	3.14	3.07	3.01	2.94	2.90	2.86	2.83	2.79	2.75	2.71
10	4.96	4.10	3.71	3.48	3.33	3.22	3.14	3.07	3.02	2.98	2.91	2.85	2.77	2.74	2.70	2.66	2.62	2.58	2.54
11	4.84	3.98	3.59	3.36	3.20	3.09	3.01	2.95	2.90	2.85	2.79	2.72	2.65	2.61	2.57	2.53	2.49	2.45	2.40
12	4.75	3.89	3.49	3.26	3.11	3.00	2.91	2.85	2.80	2.75	2.69	2.62	2.54	2.51	2.47	2.43	2.38	2.34	2.30
13	4.67	3.81	3.41	3.18	3.03	2.92	2.83	2.77	2.71	2.67	2.60	2.53	2.46	2.42	2.38	2.34	2.30	2.25	2.21
14	4.60	3.74	3.34	3.11	2.96	2.85	2.76	2.70	2.65	2.60	2.53	2.46	2.39	2.35	2.31	2.27	2.22	2.18	2.13
15	4.54	3.68	3.29	3.06	2.90	2.79	2.71	2.64	2.59	2.54	2.48	2.40	2.33	2.29	2.25	2.20	2.16	2.11	2.07
16	4.49	3.63	3.24	3.01	2.85	2.74	2.66	2.59	2.54	2.49	2.42	2.35	2.28	2.24	2.19	2.15	2.11	2.06	2.01
17	4.45	3.59	3.20	2.96	2.81	2.70	2.61	2.55	2.49	2.45	2.38	2.31	2.23	2.19	2.15	2.10	2.06	2.01	1.96
18	4.41	3.55	3.16	2.93	2.77	2.66	2.58	2.51	2.46	2.41	2.34	2.27	2.19	2.15	2.11	2.06	2.02	1.97	1.92
19	4.38	3.52	3.13	2.90	2.74	2.63	2.54	2.48	2.42	2.38	2.31	2.23	2.16	2.11	2.07	2.03	1.98	1.93	1.88
20	4.35	3.49	3.10	2.87	2.71	2.60	2.51	2.45	2.39	2.35	2.28	2.20	2.12	2.08	2.04	1.99	1.95	1.90	1.84
21	4.32	3.47	3.07	2.84	2.68	2.57	2.49	2.42	2.37	2.32	2.25	2.18	2.10	2.05	2.01	1.96	1.92	1.87	1.81
22	4.30	3.44	3.05	2.82	2.66	2.55	2.46	2.40	2.34	2.30	2.23	2.15	2.07	2.03	1.98	1.94	1.89	1.84	1.78
23	4.28	3.42	3.03	2.80	2.64	2.53	2.44	2.37	2.32	2.27	2.20	2.13	2.05	2.01	1.96	1.91	1.86	1.81	1.76
24	4.26	3.40	3.01	2.78	2.62	2.51	2.42	2.36	2.30	2.25	2.18	2.11	2.03	1.98	1.94	1.89	1.84	1.79	1.73
25	4.24	3.39	2.99	2.76	2.60	2.49	2.40	2.34	2.28	2.24	2.16	2.09	2.01	1.96	1.92	1.87	1.82	1.77	1.71
30	4.17	3.32	2.92	2.69	2.53	2.42	2.33	2.27	2.21	2.16	2.09	2.01	1.93	1.89	1.84	1.79	1.74	1.68	1.62
40	4.08	3.23	2.84	2.61	2.45	2.34	2.25	2.18	2.12	2.08	2.00	1.92	1.84	1.79	1.74	1.69	1.64	1.58	1.51
60	4.00	3.15	2.76	2.53	2.37	2.25	2.17	2.10	2.04	1.99	1.92	1.84	1.75	1.70	1.65	1.59	1.53	1.47	1.39
120	3.92	3.07	2.68	2.45	2.29	2.18	2.09	2.02	1.96	1.91	1.83	1.75	1.66	1.61	1.55	1.50	1.43	1.35	1.25
∞	3.84	3.00	2.60	2.37	2.21	2.10	2.01	1.94	1.88	1.83	1.75	1.67	1.57	1.52	1.46	1.39	1.32	1.22	1.00

DEGREES OF FREEDOM FOR DENOMINATOR

TABLE A.6 (Continued)
The F distribution (values of $F_{.01}$)*

DEGREES OF FREEDOM FOR NUMERATOR

df denom	1	2	3	4	5	6	7	8	9	10	12	15	20	24	30	40	60	120	∞
1	4,052	5,000	5,403	5,625	5,764	5,859	5,928	5,982	6,023	6,056	6,106	6,157	6,209	6,235	6,261	6,287	6,313	6,339	6,366
2	98.5	99.0	99.2	99.2	99.3	99.3	99.4	99.4	99.4	99.4	99.4	99.4	99.4	99.5	99.5	99.5	99.5	99.5	99.5
3	34.1	30.8	29.5	28.7	28.2	27.9	27.7	27.5	27.3	27.2	27.1	26.9	26.7	26.6	26.5	26.4	26.3	26.2	26.1
4	21.2	18.0	16.7	16.0	15.5	15.2	15.0	14.8	14.7	14.5	14.4	14.2	14.0	13.9	13.8	13.7	13.7	13.6	13.5
5	16.3	13.3	12.1	11.4	11.0	10.7	10.5	10.3	10.2	10.1	9.89	9.72	9.55	9.47	9.38	9.29	9.20	9.11	9.02
6	13.7	10.9	9.78	9.15	8.75	8.47	8.26	8.10	7.98	7.87	7.72	7.56	7.40	7.31	7.23	7.14	7.06	6.97	6.88
7	12.2	9.55	8.45	7.85	7.46	7.19	6.99	6.84	6.72	6.62	6.47	6.31	6.16	6.07	5.99	5.91	5.82	5.74	5.65
8	11.3	8.65	7.59	7.01	6.63	6.37	6.18	6.03	5.91	5.81	5.67	5.52	5.36	5.28	5.20	5.12	5.03	4.95	4.86
9	10.6	8.02	6.99	6.42	6.06	5.80	5.61	5.47	5.35	5.26	5.11	4.96	4.81	4.73	4.65	4.57	4.48	4.40	4.31
10	10.0	7.56	6.55	5.99	5.64	5.39	5.20	5.06	4.94	4.85	4.71	4.56	4.41	4.33	4.25	4.17	4.08	4.00	3.91
11	9.65	7.21	6.22	5.67	5.32	5.07	4.89	4.74	4.63	4.54	4.40	4.25	4.10	4.02	3.94	3.86	3.78	3.69	3.60
12	9.33	6.93	5.95	5.41	5.06	4.82	4.64	4.50	4.39	4.30	4.16	4.01	3.86	3.78	3.70	3.62	3.54	3.45	3.36
13	9.07	6.70	5.74	5.21	4.86	4.62	4.44	4.30	4.19	4.10	3.96	3.82	3.66	3.59	3.51	3.43	3.34	3.25	3.17
14	8.86	6.51	5.56	5.04	4.70	4.46	4.28	4.14	4.03	3.94	3.80	3.66	3.51	3.43	3.35	3.27	3.18	3.09	3.00
15	8.68	6.36	5.42	4.89	4.56	4.32	4.14	4.00	3.89	3.80	3.67	3.52	3.37	3.29	3.21	3.13	3.05	2.96	2.87
16	8.53	6.23	5.29	4.77	4.44	4.20	4.03	3.89	3.78	3.69	3.55	3.41	3.26	3.18	3.10	3.02	2.93	2.84	2.75
17	8.40	6.11	5.19	4.67	4.34	4.10	3.93	3.79	3.68	3.59	3.46	3.31	3.16	3.08	3.00	2.92	2.83	2.75	2.65
18	8.29	6.01	5.09	4.58	4.25	4.01	3.84	3.71	3.60	3.51	3.37	3.23	3.08	3.00	2.92	2.84	2.75	2.66	2.57
19	8.19	5.93	5.01	4.50	4.17	3.94	3.77	3.63	3.52	3.43	3.30	3.15	3.00	2.92	2.84	2.76	2.67	2.58	2.49
20	8.10	5.85	4.94	4.43	4.10	3.87	3.70	3.56	3.46	3.37	3.23	3.09	2.94	2.86	2.78	2.69	2.61	2.52	2.42
21	8.02	5.78	4.87	4.37	4.04	3.81	3.64	3.51	3.40	3.31	3.17	3.03	2.88	2.80	2.72	2.64	2.55	2.46	2.36
22	7.95	5.72	4.82	4.31	3.99	3.76	3.59	3.45	3.35	3.26	3.12	2.98	2.83	2.75	2.67	2.58	2.50	2.40	2.31
23	7.88	5.66	4.76	4.26	3.94	3.71	3.54	3.41	3.30	3.21	3.07	2.93	2.78	2.70	2.62	2.54	2.45	2.35	2.26
24	7.82	5.61	4.72	4.22	3.90	3.67	3.50	3.36	3.26	3.17	3.03	2.89	2.74	2.66	2.58	2.49	2.40	2.31	2.21
25	7.77	5.57	4.68	4.18	3.86	3.63	3.46	3.32	3.22	3.13	2.99	2.85	2.70	2.62	2.53	2.45	2.36	2.27	2.17
30	7.56	5.39	4.51	4.02	3.70	3.47	3.30	3.17	3.07	2.98	2.84	2.70	2.55	2.47	2.39	2.30	2.21	2.11	2.01
40	7.31	5.18	4.31	3.83	3.51	3.29	3.12	2.99	2.89	2.80	2.66	2.52	2.37	2.29	2.20	2.11	2.02	1.92	1.80
60	7.08	4.98	4.13	3.65	3.34	3.12	2.95	2.82	2.72	2.63	2.50	2.35	2.20	2.12	2.03	1.94	1.84	1.73	1.60
120	6.85	4.79	3.95	3.48	3.17	2.96	2.79	2.66	2.56	2.47	2.34	2.19	2.03	1.95	1.86	1.76	1.66	1.53	1.38
∞	6.63	4.61	3.78	3.32	3.02	2.80	2.64	2.51	2.41	2.32	2.18	2.04	1.88	1.79	1.70	1.59	1.47	1.32	1.00

DEGREES OF FREEDOM FOR DENOMINATOR

*This table is abridged from E. S. Pearson and H. O. Hartley (eds.), "Biometrika Tables for Statisticans," vol. 1, table 18, pp. 159 and 161, Cambridge University Press, 1962, and is reproduced with permission of the "Biometrika" trustees.

The value α is the probability that Σd^2 is less than or equal to the table value (or that r_s is greater than or equal to the table value). An observed value of Σd^2 is located in the table heading and the corresponding value of r_s and α are in the body of the table arranged by sample size N.

TABLE A.7
The rank correlation test

N								VALUES OF Σd^2						
		0	2	4	6	8	10	12	14	16	18	20	22	24
2	α	.500	1.000											
	r_s	1.00	−1.00											
3	α	.167	.500	—	.833									
	r_s	1.00	.50	—	−.50									
4	α	.042	.167	.208	.375	.458	.542	.625	.792	.833	.958	1.000		
	r_s	1.00	.80	.60	.40	.20	.00	−.20	−.40	−.60	−.80	−1.00		
5	α	.008	.042	.067	.117	.175	.225	.258	.342	.392	.475	.525	.608	.658
	r_s	1.00	.90	.80	.70	.60	.50	.40	.30	.20	.10	.00	−.10	−.20
6	α	.001	.008	.017	.029	.051	.068	.088	.121	.149	.178	.210	.249	.282
	r_s	1.00	.94	.89	.83	.77	.71	.66	.60	.54	.49	.43	.37	.31
7	α	.000	.001	.003	.006	.012	.017	.024	.033	.044	.055	.069	.083	.100
	r_s	1.00	.96	.93	.89	.86	.82	.79	.75	.71	.68	.64	.61	.57
8	α	.000	.000	.001	.001	.002	.004	.005	.008	.011	.014	.018	.023	.029
	r_s	1.00	.98	.95	.93	.90	.88	.86	.83	.81	.79	.76	.74	.71
9	α	.000	.000	.000	.000	.000	.001	.001	.002	.002	.003	.004	.005	.007
	r_s	1.00	.98	.97	.95	.93	.92	.90	.88	.87	.85	.83	.82	.80
10	α	.000	.000	.000	.000	.000	.000	.000	.000	.000	.001	.001	.001	.001
	r_s	1.00	.99	.98	.96	.95	.94	.93	.92	.90	.89	.88	.87	.86

ESSENTIALS
OF STATISTICS

N		26	28	30	32	34	36	38	40	42	44	46	48
5	α	.742	.775	.825	.883	.933	.958	.992	1.000				
	r_s	−.30	−.40	−.50	−.60	−.70	−.80	−.90	−1.00				
6	α	.329	.357	.401	.460	.500	.540	.599	.643	.671	.718	.751	.790
	r_s	.26	.20	.14	.09	.03	−.03	−.09	−.14	−.20	−.26	−.31	−.37
7	α	.118	.133	.151	.177	.200	.222	.249	.278	.297	.331	.357	.391
	r_s	.54	.50	.46	.43	.39	.36	.32	.29	.25	.21	.18	.14
8	α	.035	.042	.048	.057	.066	.076	.085	.098	.108	.122	.134	.150
	r_s	.69	.67	.64	.62	.60	.57	.55	.52	.50	.48	.45	.43
9	α	.009	.011	.013	.016	.018	.022	.025	.029	.033	.038	.043	.048
	r_s	.78	.77	.75	.73	.72	.70	.68	.67	.65	.63	.62	.60
10	α	.002	.002	.003	.004	.004	.005	.006	.007	.009	.010	.012	.013
	r_s	.84	.83	.82	.81	.79	.78	.77	.76	.75	.73	.72	.71

N		50	52	54	56	58	60	62	64	66	68	70	72	74
6	α	.822	.851	.879	.913	.932	.949	.971	.983	.992	.999	1.000		
	r_s	−.43	−.49	−.54	−.60	−.66	−.71	−.77	−.83	−.89	−.94	−1.00		
7	α	.420	.453	.482	.518	.547	.580	.609	.643	.669	.703	.723	.751	.778
	r_s	.11	.07	.04	.00	−.04	−.07	−.11	−.14	−.18	−.21	−.25	−.29	−.32
8	α	.163	.180	.195	.214	.231	.250	.268	.291	.310	.332	.352	.376	.397
	r_s	.40	.38	.36	.33	.31	.29	.26	.24	.21	.19	.17	.14	.12
9	α	.054	.060	.066	.074	.081	.089	.097	.106	.115	.125	.135	.146	.156
	r_s	.58	.57	.55	.53	.52	.50	.48	.47	.45	.43	.42	.40	.38
10	α	.015	.017	.019	.022	.024	.027	.030	.033	.037	.040	.044	.048	.052
	r_s	.70	.68	.67	.66	.65	.64	.62	.61	.60	.59	.58	.56	.55

N		76	78	80	82	84	86	88	90	92	94	96	98	100
7	α	.802	.823	.849	.867	.882	.900	.917	.931	.945	.956	.967	.976	.983
	r_s	-.36	-.39	-.43	-.46	-.50	-.54	-.57	-.61	-.64	-.68	-.71	-.75	-.76
8	α	.420	.441	.467	.488	.512	.533	.559	.580	.603	.624	.648	.668	.690
	r_s	.10	.07	.05	.02	.00	-.02	-.05	-.07	-.10	-.12	-.14	-.17	-.19
9	α	.168	.179	.193	.205	.218	.231	.243	.260	.276	.290	.307	.322	.339
	r_s	.37	.35	.33	.32	.30	.28	.27	.25	.23	.22	.20	.18	.17
10	α	.057	.062	.067	.072	.077	.083	.089	.096	.102	.109	.116	.124	.132
	r_s	.54	.53	.52	.50	.49	.48	.47	.45	.44	.43	.42	.41	.39

N		102	104	106	108	110	112	114	116	118	120	122	124	126
7	α	.988	.944	.997	.999	1.000	1.000							
	r_s	-.82	-.86	-.89	-.93	-.96	-1.00							
8	α	.709	.732	.750	.769	.786	.805	.820	.837	.850	.866	.878	.892	.902
	r_s	-.21	-.24	-.26	-.29	-.31	-.33	-.36	-.38	-.40	-.43	-.45	-.48	-.50
9	α	.354	.372	.388	.405	.422	.440	.456	.474	.491	.509	.526	.544	.560
	r_s	.15	.13	.12	.10	.08	.07	.05	.03	.02	.00	-.02	-.03	-.05
10	α	.139	.148	.156	.165	.174	.184	.193	.203	.214	.224	.235	.246	.257
	r_s	.38	.37	.36	.35	.33	.32	.31	.30	.29	.27	.26	.25	.24

N		128	130	132	134	136	138	140	142	144	146	148	150	152
8	α	.915	.924	.934	.943	.952	.958	.965	.971	.977	.982	.986	.989	.992
	r_s	-.52	-.55	-.57	-.60	-.62	-.64	-.67	-.69	-.71	-.74	-.76	-.79	-.81
9	α	.578	.595	.612	.628	.646	.661	.678	.693	.710	.724	.740	.753	.769
	r_s	-.07	-.08	-.10	-.12	-.13	-.15	-.17	-.18	-.20	-.22	-.23	-.25	-.27
10	α	.268	.280	.292	.304	.316	.328	.341	.354	.367	.379	.393	.406	.419
	r_s	.22	.21	.20	.19	.18	.16	.15	.14	.13	.12	.10	.09	.08

N		154	156	158	160	162	164	166	168	170	172	174	176	178
8	α	.995	.996	.998	.999	.999	1.000	1.000	1.000					
	r_s	-.83	-.86	-.88	-.90	-.93	-.95	-.98	-1.00					
9	α	.782	.795	.807	.821	.832	.844	.854	.865	.875	.885	.894	.903	.911
	r_s	-.28	-.30	-.32	-.33	-.35	-.37	-.38	-.40	-.42	-.43	-.45	-.47	-.48
10	α	.433	.446	.459	.473	.486	.500	.514	.527	.541	.554	.567	.581	.594
	r_s	.07	.05	.04	.03	.02	.01	-.01	-.02	-.03	-.04	-.05	-.07	-.08

N		180	182	184	186	188	190	192	194	196	198	200	202	204
9	α	.919	.926	.934	.940	.946	.952	.957	.962	.967	.971	.975	.978	.982
	r_s	-.50	-.52	-.53	-.55	-.57	-.58	-.60	-.61	-.63	-.65	-.67	-.68	-.70
10	α	.607	.621	.633	.646	.659	.672	.684	.696	.708	.720	.732	.743	.754
	r_s	-.09	-.10	-.12	-.13	-.14	-.15	-.16	-.18	-.19	-.20	-.21	-.22	-.24

N		206	208	210	212	214	216	218	220	222	224	226	228	230
9	α	.984	.987	.989	.991	.993	.995	.996	.997	.998	.998	.999	.999	1.000
	r_s	-.72	-.73	-.75	-.77	-.78	-.80	-.82	-.83	-.85	-.87	-.88	-.90	-.92
10	α	.765	.776	.786	.797	.807	.816	.826	.835	.844	.852	.861	.868	.876
	r_s	-.25	-.26	-.27	-.28	-.30	-.31	-.32	-.33	-.35	-.36	-.37	-.38	-.39

N		232	234	236	238	240	242	244	246	248	250	252	254	256
9	α	1.000	1.000	1.000	1.000	1.000								
	r_s	-.93	-.95	-.97	-.98	-1.00								
10	α	.884	.891	.898	.904	.911	.917	.923	.928	.933	.938	.943	.948	.952
	r^2	-.41	-.42	-.43	-.44	-.45	-.47	-.48	-.49	-.50	-.52	-.53	-.54	-.55

N	258	260	262	264	266	268	270	272	274	276	278	280	282
10 α r_s	.956 −.56	.960 −.58	.963 −.59	.967 −.60	.970 −.61	.973 −.62	.976 −.64	.978 −.65	.981 −.66	.983 −.67	.985 −.68	.987 −.70	.988 −.71
N	284	286	288	290	292	294	296	298	300	302	304	306	308
10 α r_s	.990 −.72	.991 −.73	.993 −.75	.994 −.76	.995 −.77	.996 −.78	.996 −.79	.997 −.81	.998 −.82	.998 −.83	.999 −.84	.999 −.85	.999 −.87
N	310	312	314	316	318	320	322	324	326	328	330		
10 α r_s	.999 −.88	1.000 −.89	1.000 −.90	1.000 −.92	1.000 −.93	1.000 −.94	1.000 −.95	1.000 −.96	1.000 −.98	1.000 −.99	1.000 −1.00		

SOURCE: W. J. Dixon and F. J. Massey, Jr., "Introduction to Statistical Analysis," 3d ed., pp. 570–574, McGraw-Hill, New York, 1969.

Chpt 6 Probability

- $E = (E \cap F) \cup (E \cap F^c)$
- $P(C \cup D) = P(C) + P(D) - P(C \cap D)$
- $P(C|D) = \dfrac{P(C \cap D)}{P(D)} = P(C)$
- independent if $P(C) P(D) = P(C \cap D)$

INDEX

Chi - Square Test   pg 204

step1N: independent   A: dependent — degrees of freedom — row — colmn

step 2 - α = significance level $df (r-1)(c-1)$

$X^2 > X_\alpha^2 [df \#]$ then yes we reject. TableA.S

step 3 - calculate expected table.

actual

	SS	MS	Σ
NT	a	D	G
AC	b	e	H
AS	c	f	I
Σ	J	K	L

expected

	SS	MS	Σ
NT	$\frac{G \cdot J}{L}$	$\frac{G \cdot K}{L}$	G
AC	$\frac{H \cdot J}{L}$		H
AS			I
Σ	J	K	L

Pg 206, 207

step 4 - Chi - square statistic $= X^2 = \sum \dfrac{(f-e)^2}{e}$

Table

| cell | f | e | f-e | $(f-e)^2$ | $\dfrac{(f-c)^2}{e}$ |

$\overline{\quad\quad}$

$X^2 = \ ?$

Chpt. 7   analysis of variance.

Step 1 — H: True means are the same
A: " aren't all the same
- sum colmns
- find mean for each colmn sum

Step 2 — Table ex p 240

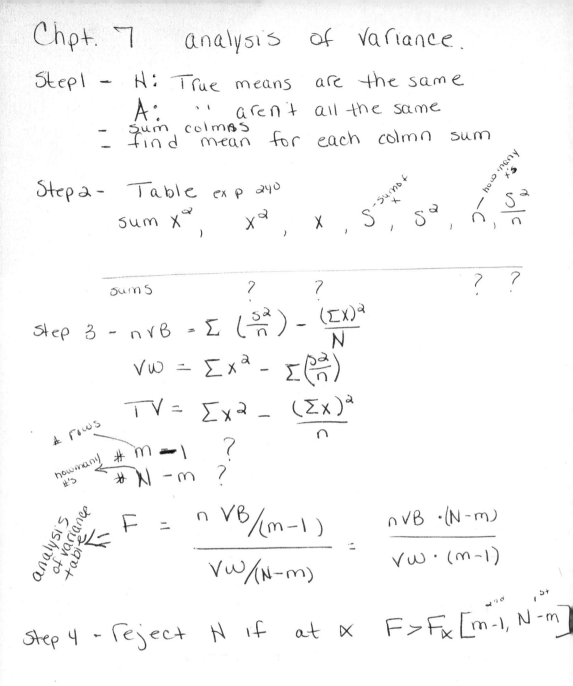

	sum x^2 ,	x^2 ,	x ,	S $^{-sumof}_{x}$,	S^2 ,	n ,	$\frac{S^2}{n}$ $^{how many}_{x's}$
sums	?	?				?	?

Step 3 — $nvB = \Sigma\left(\frac{S^2}{n}\right) - \frac{(\Sigma X)^2}{N}$

$vw = \Sigma x^2 - \Sigma\left(\frac{S^2}{n}\right)$

$TV = \Sigma x^2 - \frac{(\Sigma x)^2}{n}$

rows ⟍
how many ↙ # $m - 1$?
#'s   # $N - m$?

analysis of variance table ↙ $F = \dfrac{n\,vB/(m-1)}{vw/(N-m)} = \dfrac{nvB \cdot (N-m)}{vw \cdot (m-1)}$

Step 4 — reject N if at α $F > F_\alpha[m-1, N-m]$

Tables
A.6

vB = Variation b/t samples
vw =   "   w/in samples
TV = components of variance formula. $nvB + vw$

INDEX

Linear Regression Equation (Sec. 5.B)

$$y = a + bx$$

$$b = \frac{n\Sigma xy - (\Sigma x)(\Sigma y)}{n\Sigma x^2 - (\Sigma x)^2} \tag{5.1}$$

$$a = \frac{\Sigma y - b\Sigma x}{n} \tag{5.2}$$

Coefficient of Linear Correlation (Sec. 5.C)

$$r = \frac{n\Sigma xy - (\Sigma x)(\Sigma y)}{\sqrt{n\Sigma x^2 - (\Sigma x)^2} \ \sqrt{n\Sigma y^2 - (\Sigma y)^2}} \tag{5.3}$$

Coefficient of Linear Determination (Sec. 5.C)

$$r^2$$

t Statistic for Linearity (Sec. 5.D)

$$t = \frac{r\sqrt{n - 2}}{\sqrt{1 - r^2}} \tag{5.4}$$

$$df = n - 2$$

Arithmetic of Events (Sec. 6.A)

$$E = (E \cap F) \cup (E \cap F^c) \tag{6.1}$$

Probability of Union of Events (Sec. 6.B)

$$P(C \cup D) = P(C) + P(D) - P(C \cap D) \tag{6.2}$$

Probability of an Event (Sec. 6.B)

$$P(E) = P(E \cap F) + P(E \cap F^c) \tag{6.3}$$

Conditional Probability (Sec. 6.C)

$$P(C \mid D) = \frac{P(C \cap D)}{P(D)} \tag{6.4}$$

Probability of Intersection of Events (Sec. 6.C)

$$P(C \cap D) = P(C \mid D) \, P(D) \tag{6.5}$$

Probability of Intersection of Independent Events (Sec. 6.D)

$$P(C \cap D) = P(C) \, P(D) \tag{6.6}$$

$$P(C_1 \cap C_2 \cap \cdots \cap C_n) = P(C_1) \, P(C_2) \cdots P(C_n) \tag{6.7}$$